Lightning Protection
for
People and Property

Lightning Protection
for
People and Property

Marvin M. Frydenlund

VNR VAN NOSTRAND REINHOLD
New York

Copyright © 1993 by Van Nostrand Reinhold
Softcover reprint of the hardcover 1st edition   1993
Library of Congress Catalog Card Number 92-30334
ISBN-13: 978-1-4684-6550-1   e-ISBN-13: 978-1-4684-6548-8
DOI: 10.1007/978-1-4684-6548-8

I T P   Van Nostrand Reinhold is a division
of International Thomson Publishing. ITP
logo is a trademark under license.

Van Nostrand Reinhold
115 Fifth Avenue
New York, New York 10003

International Thomson Publishing
Berkshire House
168-173 High Holborn
London, WC1V 7AA, England

Thomas Nelson Australia
102 Dodds Street
South Melbourne 3205
Victoria, Australia

Nelson Canada
1120 Birchmount Road
Scarborough, Ontario
M1K 5G4, Canada

16   15   14   13   12   11   10   9   8   7   6   5   4   3   2   1

Library of Congress Cataloging-in-Publication Data

Frydenlund, Marvin M.
    Lightning protection for people and property / Marvin M.
Frydenlund.
        p.   cm.
    Includes index.
    ISBN-13: 978-1-4684-6550-1
    1. Lightning protection.   I. Title.
TK153.F79   1993
621.317—dc20
                                                                    92-30334
                                                                        CIP

Contents

Foreword

The word *lightning* conjures up many different thoughts, depending on who you are. For instance, an engineer might wonder, "How can I protect this building?" A computer manager might ask, "What protection is available to save my equipment?" A golf course manager wants to warn golfers about life-saving measures to take during a lightning storm. And on and on the needs go. In fact, the variety of backgrounds of people who need to understand lightning, its effects, and protection techniques ranges from the highly technical to the illiterate, creating a very difficult educational problem for the lightning protection industry.

In this book, Marvin Frydenlund presents a refreshingly new melding of the many pieces of the lightning jigsaw puzzle that are scattered throughout numerous libraries, magazine and newspaper articles, codes, standards, statistics, research, and the enormous body of information called "old wives' tales, myths, opinions, and snake oil sales literature." He has created a broad-spectrum review of the entire field of lightning that is easy to read and understand, and it will benefit anyone who wants to study and comprehend the subject.

Being an electrical engineer, I am continually reminded, as I work with other engineers, of how little knowledge exists in the engineering community with regard to lightning. It is not a course subject normally taught in college, and even the phenomenon that creates lightning is not fully understood by the scientific community. As a result, the field of lightning protection is fraught with half-truths, unsubstantiated claims for protection devices and equipment, and misapplication of protection devices and data. This generates great confusion among those who need lightning protection that really works.

Mr. Frydenlund, who has been associated with the lightning protection industry for many, many years, has rightly recognized that everyone, sooner or later, is touched by lightning and its destructive effects. Therefore, a need exists for a basic body of true knowledge to form a foundation from which to launch further, more specific studies. For example, in Chapter 13, "Surge Suppression Devices and Designs," the material presented will provide the reader with a basic understanding of surge protection devices, terminology, and their application advantages and disadvantages. It will also allow him to intelligently discuss the subject with suppressor vendors and computer equipment manufacturers to determine the correct application for his use. If this same reader later decides to consider protecting his building from lightning damage, other chapters provide the same level of necessary information to discuss the subject with an engineer and/or contractor, without being totally lost and vulnerable to accepting improper design or installation methods.

So whether you have lost a loved one, a cow, a computer, a building, a tree, or an argument on lightning theory, this book, in one convenient volume, covers all you need to be more knowledgable than most in dealing with the broad lightning field. Ben Franklin would be proud to know that his early experiments would result in such a solid body of knowledge as is presented here.

Edward A. Lobnitz, P.E.
Registered Professional Engineer
Chairman of the Board
Tilden, Lobnitz and Cooper, Inc.
Orlando, Florida

1

Lightning: Myth and Superstition

Somewhere, untold millennia ago, a hand reached out toward the glowing remains of a lightning-ignited forest blaze. Warily, the hand plucked a burning brand from the embers, carried if off, and with it started the first campfire.

It was lightning's first and perhaps last beneficial contribution to man. Aside from its usefulness as a background for theatrical drama, and a probably minor role in the greening of the world through the formation and release of atmospheric nitrate as a product of thunderstorms, lightning is to mankind a negative factor. And the more technically advanced the world becomes, the larger and more costly lightning's negative role also becomes. It could hardly have been better designed to serve as a foe of high technology.

Lightning is a complex phenomenon; a product of many variable conditions and interactions in the atmosphere. Because of this, it is probably true that no two lightning flashes are ever precisely identical. Each flash has its own peculiar mixture of properties; therefore, the effects it exerts on the object or objects it strikes are quite individual.

LIGHTNING AND MYTH

One can only speculate what the first recipients of lightning's gift of fire thought lightning was. Recorded history did not evolve until some 6,000 years ago, in the early civilizations that preceded the time of Christ.

But the writings of ancient prophets, sages, and scribes indicate that lightning was variously attributed to great gods that lived in the heavens, looking down upon the world and hurling thunderbolts at errant humans; or to the flights of giant eagles; or to the flash of the thunderbird's blinking eye.

According to writers of old, the thunderbird was not an exotic bird but one well known in the particular region—an eagle (see Fig. 1.1), a vulture, an egret, or even a pelican. Like the Bantus of Africa, some people of old believed that the thunderbird punished the evil and unclean, that the gelatinous gobs of sap from trees were the thunderbird's eggs, and that "sheet lightning" was caused by the blink of the thunderbird's fiery eye.

FIGURE 1.1   Eagles were among the giant "thunderbirds" that some of the ancient people of Greece and Rome believed ruled over earth from the skies above and sometimes blinked their fiery eyes during thunderstorms.   Drawing by Valerie Frydenlund.

Arrows of the Gods

Ancient Greeks, Romans, Persians, and Scandinavians held the view that lightning flashes were fiery shafts hurled at errant humans by gods angered by earthly transgressions. The Greeks, who once dominated the civilized world, believed that the earth was flat and circular, that Athens occupied the middle of the world, and that great gods lived in the heavens above. Each god ruled over a major feature of good or evil and meted out praise or punishment as the situation merited.

Father of all gods was the all-wise Zeus (see Fig. 1.2), ruler of the heavens and father of other gods as well as mortal heroes. Zeus punished the arrogant and errant by hurling thunderbolts at them—thunderbolts that had been invented by Minerva, goddess of wisdom, and forged by Vulcan, a lame ironsmith.

When lightning split trees or stripped them of bark, when bolts killed an animal or person, and when "hot bolts" of lightning ignited wooden buildings, the Greeks excused the incidents as erratic aim. They revered the sites of lightning strikes and occasionally built a pagan temple at a thunderstruck site.

When the Roman armies conquered Greece and seized control over much of the civilized world, they adopted Greek mythology, translating "Zeus pater" (father of gods) into Rome's "Jupiter," or "Jove" for short. The Romans continued to embellish the mythology, insisting that Jove slew monsters by striking them with lightning and that Vulcan's weapons manufacture was of such high quality that Jove presented the limping smithy with Venus as a bride.

Hindu and Norse Beliefs

The ancient Hindu religion was one of two that did not ascribe control of lightning to a single supergod. To them, thunderbolts were controlled by Indra, one of a group of gods of equal rank who passed down to the Maruts the task of scheduling storms and sending thunderbolts to earth.

The Norsemen, of what is now Scandinavia, worshiped Odin, father of gods, who had the demon fighter, Thor, hurl his great hammer at the devils of the storms that brought lightning. As Thor threw the hammer, it often struck the earth instead of the clouds in which the demons dwelled.

Apparently the Vikings were quick of eye, because they observed the fact that the illumination of lightning's return stroke, during which all the electrical current laid along the path from cloud to earth flows to ground, travels upward. In Norse mythology, this illumination was caused by Thor's hammer returning to his mighty hand.

Religion's Beliefs

The Bible's book of Job, Verse 16, cites lightning and equates back to the 11th Chapter of Genesis. During the time of the *Old Testament* prophets—of Abra-

FIGURE 1.2 Zeus, the all-wise father of the gods ruling ancient Greece, hurled thunderbolts to punish wrongdoers and the arrogant. Zeus was followed by Jupiter, chief god of the succeeding Roman Empire.   Drawing by Valerie Frydenlund.

ham, Moses, and David—appearances of God were accompanied by thunder and lightning.

The Moslem Koran states, "He launcheth thunderbolts and smiteth with them whome He will while they dispute concerning Allah, and He is mighty in wrath."

A Philosopher's Theory

The ancient philosopher Aristotle, fourth-century pupil of Plato, theorized that as air moved on the ground, an "exhalation" was created that rose and collected in a layer between the earth and the moon. During hot weather, Aristotle said, this exhalation created lightning, thunder, and wind. When the weather was cool and the air damp, clouds were formed, and rain, hail, or snow was created.

As he explained his theory, Aristotle cited a hot, dry layer that lay between the sun and the moon, which he referred to as a "region of fire." This created a belief that lightning spurted from a region of fire near the moon.

The Rise and Fall of Augury

The fifth great myth of antiquity was the notion that thunder and lightning were omens foretelling the coming of good or evil. Ancient Greeks faced northward as they listened for the sounds of fortune. If thunder rumbled on the right, that was a good sign. If it thundered to the left, that was a bad omen, and it was best to shelve any great plans until a change in the weather occurred.

Storms in the area tended to travel from west to east. Thus, the Greeks, who faced north to listen to the sounds of prophesy, heard the approaching storm on the left and the departing storm on the right.

When the conquering Romans adopted the practice of listening for the sounds of fortune, they faced south rather than north to listen to the counsel of the thundergods. So doing, they switched the signs of good and evil.

By 300 B.C., listening for omens had become quite a complex business. Directions taken by shooting stars, and even by the flights of birds, were taken into account. A College of Augurs, whose members became important personages that took themselves increasingly seriously, was established.

Eventually power corrupted, and augurs began to report signs that weren't there in order to gain their political ends. The Roman people eventually had enough of this nonsense, and augury died of its own excesses.

The Magic of Wood

Perhaps the most durable of ancient lightning related myths was the notion that there was a magic in wood that could prevent lightning from striking. Some people thought that a chunk of wood from a tree that had been struck by lightning acquired a sacredness, and they refrained from burning it for fear of Heavenly reprisal. Others believed the opposite—that burning a piece of lightning-struck wood in the fireplace would act as a sort of immunization; that, like a vaccination, the charred chunk would prevent a full-scale onslaught by lightning.

A modified version of the magic wood theory persisted in Europe for centuries and was finally put to rest during the nineteenth century. Englishmen and Germans believed in retaining the traditional Yule log, a length of oak, so that they could toss it into the fireplace and ignite it to ward off lightning. The French and Flemish, on the other hand, believed that a chunk from a lightning-struck tree could best ward off lightning from a position under the bed.

The Emperor's Defense

Ancient Romans took a somewhat different approach. Pliny the Elder, a Roman naturalist and writer of the second generation after Christ, advised that laurel bushes could prevent lightning from striking. This so impressed then Roman emperor Tiberius that he took to keeping a laurel wreath handy and would drape it around his neck during thunderstorms.

Laurel bushes, not being very tall and prominent, are not as likely to be struck by lightning as are larger trees. One may surmise that the observant emperor based his theory on that fact. Later, the idea got around Rome that if laurel bushes could ward off the harm of thunder and lightning, so could thorn bushes. And so, during thunderstorms, many Romans strode around Rome with thorns in their togas.

Bells Versus Demons

Tiberius' notion that a laurel wreath would protect him from lightning's harmful intent was quaint enough. But it was followed by a practice that, while also quaint, was also dangerous. This was the practice of ringing church bells to confront, and subdue with holy noise, the Satanic spirits whose evil doings were blamed for the crashing of thunder and flashing of lightning.

History is vague on how the notion took hold. One theory is that, at first, church bells were rung to call parishioners together in prayer, that they might be spared from harm by lightning. In the course of time, the original purpose was forgotten, and it came to be believed that it was the pealing of the bells rather than the power of prayer that provided the protection.

Another theory was that the "sympathetic magic" wrought by the holy music of the bells could overcome the evil noise of thunder. In either event, from the thirteenth century through the eighteenth century and beyond, church sextons rushed to their belfries to jangle the supposedly hallowed bells as thunderclouds scudded toward a community. People believed the admonition of St. Thomas Aquinas, Italian scholastic philosopher of the thirteenth century, who wrote, " . . . the tones of the consecrated metal repel the demons and arrest storms and lightning."

Danger in the Steeple

Statistics show that the local steeple was a distinctly unhealthy place to be (see Fig. 1.3). It was built to reach toward Heaven and topped a church that, more often than not, was built on the highest spot in the neighborhood in order to better reach toward Heaven. A 1784 book printed in Germany reported that during a 33-year period, lightning ravaged 386 churches in Europe, and in 103 of them an unfortunate sexton died while dutifully pumping a rope to jangle a "consecrated" bell.

FIGURE 1.3 For five centuries, from the thirteenth through the nineteenth and into the early twentieth, the church bells of Europe clanged their supposedly holy music to drive away the "demons" of the thunderstorms. Sometimes, however, lightning struck belfries as the bells were clanging.

The practice of ringing church bells to dissipate lightning was encouraged by the Abbe Nollet, a leading French churchman. Nollet did not believe in Benjamin Franklin's lightning rods, calling their use "dangerous." He argued, "Bells, by virtue of their benediction, should scatter the thunderstorms and preserve us from strokes of lightning."

Franklin found it difficult to understand why Nollet did not advocate the use of lightning rods. He noted, "Lightning seems to strike steeples by choice, and that at the very time the bells are ringing; yet they still continue to bless the new bells, and jangle the old ones whenever it thunders."

Franklin also observed, " . . . among the great numbers of houses furnished with iron rods in North America, not one so guarded has been materially hurt by their means; while a number of houses, churches, barns, ships, &c. in different places, unprovided with rods, have been struck and greatly damaged, demolished or burnt."

Do the Bells Still Toll?

The rumble of thunder and the strident response of church bells played a musical duet that became so familiar in the storm-prone Tyronese Alps of Austria and Italy that it was called a "local institution." Even in highly civilized England, a church bell challenged the supposed demons of the storm at least once during the twentieth century. An English authority on bells, J.J. Raven, wrote in 1906 that there very likely was a rational explanation for the power bells had to "dispel the harm of thunder."

Perhaps the bells still toll somewhere in mankind's most remote areas in response to the challenging boom of thunder. But the last church bell known to have pealed for that purpose rolled its music against the crags of the Austrian Alps just before the eruption of World War I in 1914.

2

Benjamin Franklin's "Perfect Invention"

On a sultry day in June, 1752, 46-year-old editor and publisher Benjamin Franklin, with his 20-year-old son, Thomas Folger, went into a field near Philadelphia and lofted a kite into the sky. Franklin let out the string to come as close as he could to the scudding thunderclouds.

The kite was a homemade affair, but apparently it flew well enough. Franklin had fashioned it himself, selecting for the sail a large handkerchief of silk, which is somewhat resistant to rain. He stretched the cloth over two crossed cedar sticks. Projecting from the top of the vertical stave was a wire about a foot long which had been sharpened to a needle point.

As the kite rose, Franklin played out the twine and stepped into the doorway of a nearby shed. He wanted to keep the rain away from a short, silk ribbon he had attached to the end of the twine.

For a long moment the kite soared and dipped on the gusting wind, its uneventful flight a tribute not to scientific genius but to Franklin's craftsman's skill. It is easy to visualize today that, as the kite continued to fly like any other kite and nothing out of the ordinary happened, Franklin fidgeted a bit impatiently. Years of dedicated research and Franklin's budding reputation were blowing in the wind.

But then the strands of the twine began to stand out and bristle like the hairs in the scruff of an angry dog. Franklin raised a knuckle toward the dangling key and slowly brought his hand closer to the metal.

A tiny spark leaped between the key and the hand. Franklin was, no doubt, deeply elated. He had struck scientific truth (see Fig. 2.1).

The historic kite was flown during the sixth of ten years that Benjamin Franklin devoted to a study of electrical phenomena, invention of the lightning rod, and development of lightning protection principles. It was a decade sandwiched between a publishing career that made him wealthy and the three decades of service to his country that did the most to make him famous.

The celebrated flight of Franklin's kite, witnessed only by Thomas Folger Franklin, was impromptu. Although Franklin's daily journals were lost, it can be envisioned that he quite hurriedly assembled the kite so that he could fly it, spur-of-the-moment, on the arrival of a sudden thunderstorm.

Franklin had meant to conduct a grander test on a far more complex scale, following a procedure he had proposed to scientific colleagues. He called this procedure "the sentry box experiment." He had planned to have an apparatus erected atop one of the spires of a large new church, the Philadelphia Church of Christ.

FIGURE 2.1   Benjamin Franklin's famous kite flight was an impromptu substitute for his planned churchtop "sentry box experiment," but it proved that lightning is electrical current. Franklin then went on to announce his invention of lightning rods.

But then, as today, builders tended to dillydally. Eventually, Franklin grew impatient and flew his kite as an impromptu substitute.

Crude though it was, the experiment brought down from the skies one of the most important truths ever divulged by nature to investigative science—the fact that lightning's properties are electrical. Franklin later listed lightning current as identical on all counts, except magnitude, to the intriguing "subtile fluid" that he and his experimental colleagues had been producing with elementary generators (static machines) and collecting in the first crude condensers (Leyden jars).

The historical trickle of "electrical fluid" probably tickled Franklin's knuckle as gently, or possibly even more gently, than electrical shocks you may have experienced more than once—touching a door knob after crossing a plush carpet in winter, attaching jumper wires to an auto battery, crossing an electric stock fence, or grasping a shorted appliance. Later, Franklin revealed that, as the kite twine became wet in the splattering rain, he drew stronger and stronger sparks from the key and led the electrical fluid into a Leyden jar.

Ironically, Franklin did not know, as he reeled in his kite and trudged home with his paraphernalia, that he was being toasted in Paris at that very moment for the same scientific breakthrough he had just accomplished. Two Frenchmen had coaxed electric current out of thundercells, just as Franklin had theorized that it could be done. They had erected the "Sentry Box" apparatus just as he had proposed.

May 10, 1752, is recognized by the scientific community as the date of first success for the Philadelphia Experiment. On that day, M. D'Alibard drew sparks from a thundercloud through a 40-foot high, pointed iron rod, erected per Franklin's suggestions at Marly, near Paris (see Fig. 2.2). A week later, a thundercloud passed over the house of M. Delor in Paris, and a 99-foot high iron rod on the house gave off sparks an inch long.

Throughout the summer of 1752, several English members of Europe's scientific community also successfully conducted the "Philadelphia Experiment." To the everlasting credit of D'Alibard and the others, full honors for the breakthrough were given to the American. And even before the news of the European successes finally reached him via sailing ship and stagecoach, Franklin was being hailed throughout Europe as a modern Prometheus—the Titan of Greek mythology who stole the secrets of fire from the heavens and taught them to man.

Later, English theologian and chemist Joseph Priestly, discoverer of oxygen, described Franklin's findings as "the greatest, perhaps, that have been made in the whole compass of philosophy since the time of Sir Isaac Newton." But the Philadelphian had detractors as well as admirers. Some critics were motivated by envy; others by frank differences of scientific opinion.

Leading French churchman and philosopher Abbé Nollet stung Franklin with a critical paper published in the memoirs of the Royal French Academy of Sciences, calling the use of metal conductors outside a building useless or dangerous.

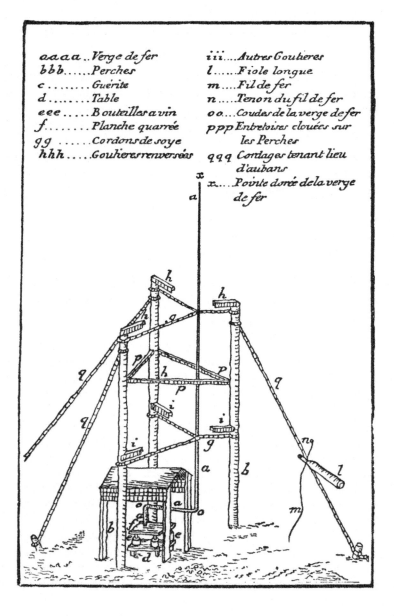

FIGURE 2.2   On May 10, 1752, this apparatus erected by Frenchman Thomas-Francois D'Alibard as Franklin recommended produced sparks during a thunderstorm, proving Franklin's cloud electrification theory.   Drawing courtesy of the American Philosophical Society.

The envious Abbé so forcibly opposed the use of lightning rods that Franklin remarked in a letter to a friend, "Nollet is applying himself to the superstitious prejudices of the populace, which I think unworthy of a philosopher."

Benjamin Wilson of England led opposition to Franklin in a controversy over whether lightning rods should be pointed or blunt. To Franklin then, and to the greater part of the lightning protection industry that he fathered, the electrically attractive properties of a sharp point constituted the very essence of lightning protection. Wilson thought, on the other hand, that pointed rods "may promote the very mischief we mean to prevent." He favored blunt points or knobs, which he believed would repel lightning, and proposed that they be placed below the roofline.

Wilson, a Fellow of the prestigious Royal Society, had access to the best-equipped laboratories of the day and enjoyed a tested scientific reputation. There is no reason to doubt that he was among the abler men of mid-eighteenth century science.

Franklin, across the sea in colonial Philadelphia, was far removed in time and place from the world's centers of science and philosophy. Facilities available to him were crude. His critics then and since have called him an amateur and a dabbler.

Lightning itself was the first to authoritatively judge Franklin's and Wilson's views. History has since confirmed that Wilson's concepts were mostly wrong and Franklin's fundamentally correct. Benjamin Wilson's tragedy was that he began with a mistaken assumption—that the attractive influence of a pointed lightning rod might cause harm. He doubted that the stroke could be conducted safely to ground. This conviction was shared and nurtured by others, creating the notion that the attractive action ought to be reversed. Insulating glass balls were fitted to rods in the hope that they would repel lightning strokes.

Lightning tested such "repellant" glass balls after they had been installed on the steeple of Christ Church in Doncaster, England. They were found wanting: the steeple was demolished.

Benjamin Franklin's triumphs were built on a correct first assumption, that "electricity is an element diffused among, and attracted by other matter," and that all bodies have a common stock of electricity that can be added to or subtracted from, causing them to be negatively or positively charged.

Building a thesis on that solid foundation, Franklin stacked experimental findings one atop the other, proving and keeping, or disproving and discarding, his own ideas as he went along. In the end, he had built a philosophical–scientific framework of such breadth, scope, and accuracy that it is, in essence, up-to-date to this day.

The following narrative presents the chronology of Franklin's work in lightning research and invention.

1746

Franklin and three friends—Thomas Hopkinson, Ebenezer Kinnersley, and Philip Syng—began their investigations. They had been intrigued by the scientific meaning behind the results of a current fad: drawing room demonstrations in which experimenters made their friends jump with charges of frictional electricity. They were spurred on also by the invention that year, at Leyden, in the Netherlands, of a condenser for static electricity. This came to be called the Leyden jar (see Fig. 2.3).

1747

Franklin, leader of the small group called the *Library Company of Philadelphia*, wrote, "I never was before engaged in any study which so totally engrossed my attention." In one of a succession of letters he sent to his friend Peter Collinson, of London (letters that later proved to be the only continuing record of his pio-

FIGURE 2.3   Ben Franklin identified lightning as electricity after years of study. During this time, he produced frictional electricity, which he stored in a Leyden jar and then discharged. Comparing the sparks produced in his laboratory and those brought down on his kite string, which he also stored in a Leyden jar, he found them to be one and the same.

neering scientific work), Franklin described his conception of electricity. He believed it to be a single, "common stock," electron fluid. He set three goals: (1) Use of the one-fluid theory in electrical experiments, (2) proof of the identity of thunderstorm and laboratory electricity, and (3) finding a practical use for the new knowledge.

In May of 1747, Franklin wrote his first scientific communication to Collinson:

> If you present the point (of a long grounded needle) in the dark, you will see sometimes at a foot distance or more, a light gather upon it like that of a fire-fly or glow-worm; the less sharp the point the nearer you must bring it to observe the light; and at whatever distance you see the light, you may draw off the electric fire.

1748

Apparently, this was a year of frustration and trial. Franklin revealed that the Library Company of Philadelphia had become impatient: "We are chagrined a little that we have been hitherto unable to produce nothing in this way of use to mankind."

1749

This was a good year for Franklin. He had described himself as "intrigued" in 1746, "totally engrossed" in 1747, and "chagrined" in 1748. His 1749 letters to Collinson indicated that he had shaken off the pessimism of the year before and had become hopeful of progress. And, sure enough, he did make greater strides that year than he knew at the time. Probably the most important of all Franklin's experiments was conducted this third year, leading directly to the celebrated Philadelphia Experiment.

Franklin built a tubular conductor of pasteboard covered with Dutch metal, 10 feet long and 1 foot in diameter, and hung it vertically from the ceiling. He charged its surface with his static machine and found that it gave off a 2-inch spark when a grounded blunt rod was brought to that distance from it. But the conductor could be discharged silently, he found, when a grounded bodkin was held a full foot away.

Presumably, this bodkin was a large, sharp needle used for making holes in cloth. Comparing its action with that of a blunt rod, Franklin demonstrated his unique ability to unclutter his mind, setting aside existing notions and reasoning anew. He insulated the discharging point and wrote, "If a person holding the point stands upon wax, he will be electrified by receiving the fire at that distance."

Franklin's minute book for November 7, 1749, listed similarities between laboratory-induced "electric fluid" and lightning. He concluded, "The electric fluid is attracted by points. We do not know if this property is in lightning. But since they agree in all the particulars wherein we can already compare them, is it not probable that they agree also in this?"

Later in 1749, Franklin charged two side-by-side gun barrels with his static machine and made another observation in a succession of comments that proved that his mind moved logically—not "by good luck and frog leaps," as some detractors have charged. He wrote:

> If two gun barrels will strike at two inches distance, and make a loud snap, to what great a distance may 10,000 acres of electrified cloud strike and give its fire and how loud must be that crack?

The bulk of Franklin's notations were lost, but in all probability, as early as 1749, he was able to interpolate, on the spot, from his lab experiments to a natural thunderstorm. He saw his suspended pasteboard-and-metal cylinder as a thundercloud, his static machine as storm currents, and his bodkin as a sharply pointed lightning rod rising from the ground.

He convinced himself by laboratory-level evidence that the pointed rod would discharge a cloud (his metal and pasteboard tube) without thunder (the snap from a 2-inch spark that occurred when a blunt lightning rod was used). Blunt rods, on the other hand, would not discharge a cloud silently, Franklin concluded.

In these first observations, Franklin was partly right and partly wrong. He was correct, as he and others proved three years later, that static sparks and lightning strokes are the same except for size. He was also correct when he proposed that, as in the laboratory, lightning discharges could be intercepted and harmlessly grounded by lightning rod systems.

Franklin was incorrect, however, in one major surmise—when he concluded that sharp-pointed lightning rods would silently discharge thunderclouds. The American lightning rod industry made that claim during its early years, then for a time equivocated, citing it as an alternate function, and finally dropped the claim altogether.

Another attempt at "silent discharge" was made by installing a large number of points at rather low levels. This also inhibits the electrical separation that results in a lightning discharge, but as critics point out, groves or forests of trees, with their multiplicity of points, are struck as often as other lightning targets.

It has been shown that Franklin also erred when he concluded, on the basis of his laboratory experiments, that lightning rods ought to be sharply pointed. One of the theories cited is that a lightning discharge is too huge to differentiate between a sharp rod and a blunt one. The lightning protection industry, wedded to sharply pointed rods for generations, has been loath to accept advice from atmospheric scientists who, by and large, have advised the use of blunt rods.

However, widely recognized authority Charles B. Moore, of Langmuir Laboratory, Socorro, New Mexico, compared sharp and blunt points atop poles in atmospheric tests. He reported that the latter were much more capable of attracting downcoming lightning flashes than their sharp counterparts. He explained that a sharp rod will readily release ions to the atmosphere, thus surrounding itself with an inhibiting ion cloud, while a blunt rod will hold a bound charge until the attraction of a downcoming streamer becomes quite intense. At some point, depending on the electrical magnitude and distance separating the two charge centers, the attraction between them will overcome the resistance of the nonconductive air, and a ground based streamer will shoot upward to meet the downcoming electron stream.

In any case, the American lightning protection industry, during the last few years, has been installing an ever larger percentage of air terminals with broad hemispherical tips. Some British and other lightning protection systems forego Franklin's vertical lightning rods in favor of horizontal conductors. The danger of ignition of flammable material on the roof is met by propping the conductors on standards.

As "Franklin rod" installations grew in number and longevity, it became apparent to Franklin that they were not preventing lightning strikes. From then on, he was careful to stress what is still recognized as the single function of lightning protection—to intercept and harmlessly ground any discharge.

1750

In 1750, Franklin made his two famous proposals: (1) for the Philadelphia experiment, and (2) for installation of lightning rods. He wrote to Collinson:

> To determine whether clouds that contain lightning are electrified or not, I would prepare an experiment to be tried when it may be done conveniently. On the top of some high tower or steeple, place a kind of sentry-box, big enough to contain a man, and an electrical stand. From the middle of the stand let an iron rod rise and pass bending out of the door and then upright 20 or 30 feet, pointed very sharp at the end. If the electrical stand be kept clean and dry, a man standing on it when such clouds are passing low might be electrified and afford sparks, the rod drawing fire to him from a cloud. If any danger to the man should be apprehended (though I think there could be none) let him stand on the floor of his box and now and then bring near to the rod, the loop of a wire that has one end fastened to the leads (earth), he holding it by a wax handle; so the sparks, if the rod is electrified, will strike from the rod to the wire, and not affect him.

Franklin proposed lightning rod installations in 1750 after making three deductions based on his laboratory experiments:

1. If thunderstorm and laboratory electricity are the same, so also are laboratory and storm conditions, dimensions being appropriately scaled down.
2. The attraction between a suspended conductor and a blunt rod before the discharge is like the downward movement of a cloud toward a hill or high building, which would pull the cloud to within its "striking distance."
3. Since a needle near to or on the blunt punch discharges the conductor silently, the same thing could be made to happen to a thundercloud which, even if it "come nigh enough to strike, yet being deprived of its fire, it cannot."

Then he continued with two separate communications:

I say, if these things are so, may not the knowledge of this power of points be of use to mankind in preserving houses, churches, ships, etc. from the stroke of lightning, by directing us to fix on the highest parts of these edifices, upright rods of iron made sharp as a needle, and gilt to prevent rusting, and from the foot of these rods a wire down the outside of the building into the ground, or down one of the shrouds of a ship and down her side until it reaches the water? Would not these pointed rods probably draw the electric fire silently out of a cloud before it came nigh enough to strike, and thereby secure us from that most sudden and terrible mischief?

If . . . on the tops of weathercocks, vanes or spindles of churches, spires or masts, there should be put a rod of iron 8 or 10 feet in length, sharpened gradually to a point like a needle, and gilt to prevent rusting, or divided into a number of points, which would be better—the electrical fire would, I think, be drawn out of a cloud silently, before it could come near enough to strike; only a light would be seen at the point, like the sailor's corpusante.

1751

It was in this fifth year of scientific study that Franklin's thinking in regard to silent discharge by lightning rods was adjusted. He pondered a description of a storm at sea where upward streamers of "electric fire," variously called "sailor's corpusantes" and "St. Elmo's fire," "burnt like very large torches" above a ship's masts, and yet the ship was struck by lightning. Obviously, the high masts of the ship had not silently discharged the thundercloud above, and Franklin retreated, but only part way, from his original view. He suggested that there would have been no stroke "had there been a good wire communication from the spintle-heads to the sea . . . ." But then he added the key "if," which he later never forgot: " . . . or if a stroke, the wire would have conducted it all onto the sea without damage to the ship."

1752

This was Franklin's first year of worldwide glory: the year he flew his kite and created a legend; the year that scientists in France and England tested his theses through various versions of the Philadelphia Experiment; the year in which he became Prometheus reincarnated.

1753

Franklin's fame, his hero status among common folk, and his towering stature among scientific contemporaries did not come to him through genius alone but also through certain human qualities. He was humble—as modest about his scientific achievements as he was later, when he helped Thomas Jefferson draft the Declaration of Independence and then declined to share the credit.

In the 1753 issue of *Poor Richard's Almanack*, Franklin squeezed his public announcement of the invention of lightning rods* between announcements of meetings of the Mayor's Court and Quakers. Franklin claimed no credit for himself:

> It has pleased God in His goodness to mankind, at length to discover to them the means of securing their habitations and other buildings from mischief by thunder and lightning. The method is this: provide a small iron rod (it may be made of rod- iron used by the nailers) but of such length, that one end being three or four feet in the moist ground, the other may be six or eight feet above the highest part of the building. To the upper end of the rod fasten about a foot of brass wire, the size of a common knitting needle, sharpened to a fine point; the rod may be secured to the house by a few small staples. If the house or barn be long, there may be a rod or point at each end, and a middling wire along the ridge from one to the other. A house thus furnished will not be damaged by lightning, it being attracted by the points, and passing through the metal into the ground without hurting any thing. Vessels also, having a sharp-pointed rod fix'd on the top of their masts, with a wire from the foot of the rod reaching down, round one of the shrouds, to the water, will not be hurt by lightning.

Lightning rods existed in Franklin's mind several years before he "invented" them. He was confident before he lofted his kite that a charge of static electricity would spark between his hand and the metal key. Obviously, Benjamin Franklin richly earned the accolades heaped upon him by other lightning protection experts, then and since.

B. F. J. Schonland, of South Africa, wrote that Franklin " . . . was a natural philosopher of very great stature, and an amateur, like Cavendish, in the best sense of the word."

Harald Norinder, of Sweden, observed, "He showed a rare capacity for observation and a logical ability in drawing conclusions in many fields."

I. Bernard Cohen, of America's Harvard University, noted, "In Franklin's willingness to rise above the superstitions of his age, and particularily in the ease with which he ignored the possibility of a Prometheus-like fate and the wrath of the father's rod, we see him as an emancipated spirit and a herald of our modern age."

And R. H. Golde, of England, some of whose predecessors fought Franklin in the points-versus-blunts controversy, quite recently had this to say: "Disciplined imagination: this is indeed the attribute which seems best to explain the outstanding success achieved in little more than a decade's work which Benjamin Franklin, the printer and diplomatist, could devote to scientific enquiry."

It was lightning itself, however, that paid Franklin the highest compliment, one day in the year 1762. Among houses in Philadelphia which had been protected by Franklin rods was one belonging to a Mr. West. When a bolt struck it, the house was saved, the lightning current " . . . being attracted by the points, and passing

---

How to Secure Houses &c, From Lightning.

through the metal . . . " as Franklin had predicted seven years before (see Fig. 2.4).

To Franklin's further credit, he was not dismayed when he found that the bolt did not quite pass "into the Ground without hurting any Thing." He inspected mi-

FIGURE 2.4 In 1762, Franklin inspected one of his lightning rods, which had been installed on a house belonging to a Mr. West, who had complained of lightning damage to the stones of his house foundation. Franklin noted that the rod should have been brought away from the house and " . . . sunk deeper so as to come to earth moister."

FIGURE 2.5   Three Franklin ideas that had less merit were a bed to be suspended by silken ropes to insulate the sleeper should lightning strike his house; "Franklin wires" for ladies' hats, which certainly would make their wearers more likely to be zapped by "step voltage"; and, finally, the notion that a lightning rod would "silently draw electric fire from the cloud"—in other words, prevent lightning from striking.

nor damage to the stone foundation of Mr. West's house where the rod crossed it and phrased what much later became lightning protection code stipulations.

Franklin correctly laid the damage to inadequate grounding. The quarter-inch thickness of the rod seemed adequate, he observed, but then prophetically added, "larger may be sometimes necessary."

He was more positive about grounding depth: "The rod should have been sunk deeper, till it came to earth moister and therefore apter to receive and dissipate the electric fluid."

Since that time, lightning protection code writers and lightning protection industry members around the world have seen to it that Ben Franklin's advice has been followed (however, see Fig. 2.5).

3

Thundercells
and Thunderstorms

Most lightning activity in North America is the result of a separation of electrical charges due to air turbulence created when a summertime cold front from the north collides with a warm air mass sweeping up from the south. The "frontal" or "squall line" storms thus created are large-scale events whose turbulent fronts may extend several hundred miles, across several states.

Two other types of convective electrical storms are prevalent in tropical and mountainous regions the world. Common in certain areas of the United States and Canada, they are (1) localized "orographic" or "mountain" storms, and (2) "lake-effect" storms.

Less common and quite different are thunderstorms that occur in temperature extremes. Lightning sometimes flashes during winter snowstorms, particularly in the Great Lakes region. And lightning will flash infrequently during the summertime from shallow clouds whose tops do not reach freezing levels.

SQUALL-LINE STORMS

Thunderclouds are cellular in nature, and a broad squall line may consist of dozens or even hundreds of cells in varying stages of formation, maturity, and decay. A thundercloud is a huge mass of electrical energy, of perhaps 1,000 coulombs, distributed over a space averaging as much as 30 cubic miles.

In a typical summertime frontal storm, dense, heavy, cold air from the north invades a lighter mass of moist warm air (see Fig. 3.1). Pushed higher and higher upward as the cold mass continues to pour in and displace it, the lighter air cools as it rises and begins to precipitate raindrops. The falling drops are carried back up again and again. Reaching freezing heights, the raindrops turn to snow and eventually to ice crystals.

Small cumulous clouds begin to form as the convective action described above begins. These clouds are the seedlings of thundercells that grow vigorously upward at speeds of 100 feet per second or more. As the top of a growing cell reaches a height of about 28,000 feet, lightning begins to flash (see Fig. 3.2).

An average frontal thundercell will reach a ceiling of 37,000 feet, depending on factors related to geographical location. The minimum diameter of such a cell is about 300 feet. "Giant" thundercells reach heights of 60,000 feet or more and mushroom out to diameters as great as 30,000 feet.

The durations of squall-line storms vary from hours to a day or more. During their lifetimes, such storms remain stationary for several hours, keeping critical operations in their expected paths of travel (like aircraft baggage loading and unloading) in a state of suspense.

Many thundercells form, mature and decay across a stationary or a moving squall line. The cells have normal lifetimes ranging from 30 to 60 minutes.

LAKE-EFFECT STORMS

A common example of a localized convective thunderstorm is created when warm, moist air rising from a sun-drenched beach is invaded and lifted skyward by denser, heavier air pouring in off the cool waters of an ocean or a large lake. Such "lake effect" storms treat cities and villages ringing the Great Lakes with a variety of sudden squalls during spring, summer, fall and winter.

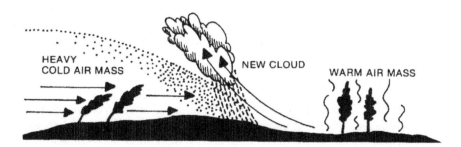

FIGURE 3.1   Most lightning storms in the United States are produced when a cold front from the north invades a warm air mass. Turbulence created as the cold front uplifts and displaces the lighter warm air results in a separation of atmospheric electric charges.

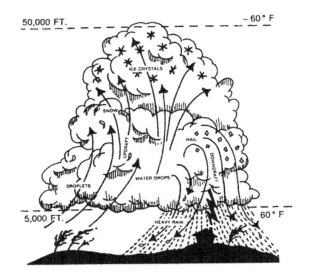

FIGURE 3.2   This thundercloud has reached its mature stage, and a downdraft has developed on the cloud's lee side, producing a sudden drop in temperature, wind gusts, rain (and perhaps hail), and flashes of lightning.

Because large bodies of water maintain much more even temperatures than do land masses, particularly sandy beach areas that can become sun baked in a matter of hours, water and land sometimes change roles in the convective process that breeds lightning (see Fig. 3.3). For example, a spring wind blowing over still-frigid water will uplift the warm air over a sun-baked area of sand dunes and "blow up a storm." In the fall, the process may be reversed as cool air from a land mass

FIGURE 3.3   A "lake effect" thunderhead is produced when cool air from a large body of water displaces air warmed by a sun-drenched land mass. Resulting convective storms usually occur in the afternoon or evening after the sun has raised the temperature of the land well above that of the adjacent water.

on a cloudless day invades moist, warm air over lake water that is still carrying some of the summer's heat.

OROGRAPHIC THUNDERSTORMS

Summertime thunderstorms are highly predictable events that sometimes occur daily in areas consisting of high mountain peaks. Small, white, thinly vaporous clouds begin to form around a mountain peak's midsection or higher in mid morning. These young cumulous clouds are the results of a replacement of air rising from sun-warmed lower slopes by cooler, heavier air drifting downward (see Fig. 3.4).

As the day wears on, the clouds grow in height and size until they ring the peak of a high mountain or shroud the top of a lower one. At some point, from near-noon to mid afternoon, the clouds begin to darken and grow faster, perhaps totally obscuring the mountain's upper levels, and lightning begins to flash.

Orographic storms cause many forest fires and are severe threats to houses and other structures built on mountain slopes. Mountain climbers are sometimes caught unexpectedly by lightning flashes as benign-looking orographic clouds quickly darken into dangerous thundercells. Scientists point out, however, that mountains provide "heat leaks" to the atmosphere which prohibit orographic thunderstorms from attaining great size.

A final example of areas prone to experience convective storms might be a valley or prairie city, its concrete and blacktop areas baking in the afternoon sun, ringed by cool forested hills. Air convection, the appearance and growth of cumulous clouds, gusting winds, and separation of electric charges capped by a late afternoon or evening thunder squall are possible events for such a city.

FIGURE 3.4   A "mountain storm" may be produced when cool air from a higher elevation drifts down a slope to replace air warmed by the sun. Or, cool air over the cold waters of a mountain lake may invade warmer air on a lower slope.

The addition of lake-effect and orographic thunderstorms, ripening around midday, helps to explain why most lightning destruction, fatalities, and injuries occur between noon and midnight, with 6:00 P.M. being the peak lightning hour.

THE CLOUD ELECTRIFICATION PROCESS

A fascinating mystery that nature still conceals from questing scientists and others is the process by which clouds become electrified. It is widely accepted that a typical thundercell is created when a complex interaction of raindrops, snowflakes, ice crystals, charged particles, and energy exchanges takes place. But many different theories have been advanced to explain how lightning is created by the resulting tumult.

"Breaking drop," "ionization," and "influence" theories are among those that propose electrification to be a result of differences in the ion-carrying characteristics of raindrops that vary in size and fall rate.

The late Ralph H. Lee, an electrical engineer and IEEE Fellow who worked with the author on several occasions during seminars and in loss investigations, was a proponent of the "breaking drop" theory of cloud electrification and lightning flash development (see Fig. 3.5).

According to Mr. Lee:

Cumulus clouds are the upper ends of gently rising columns of air. These columns, coming from near the earth, contain amounts of moisture that depend on the humidity near the ground. Where these columns, cooling as they rise, reach the level at which the relative humidity increases to 100 percent, the moisture condenses. At earth's surface, it would be called fog, since it consists simply of widely dispersed subdrops of water, floating in the air.

When the velocity of the rising air column exceeds the normal 'cumulus' status, the height of the condensed cloud increases as well. The separate small droplets have more

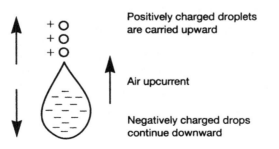

FIGURE 3.5   This illustration is based on the "breaking drop" theory of cloud electrification. The small, broken-off droplets are said to float in the upper regions of a forming thundercloud, while the heavier, negatively charged drops mass in the lower regions of the cloud.

opportunity to impinge on each other and coalesce, becoming larger droplets. When they reach appreciable size, they become so heavy that they cease to float and tend to fall downward, through the clouds, and out the bottom as rain.

With even higher vertical velocities, according to one theory, the falling drops tend to develop into teardrop shape. The drops develop separation of charge, with the lower, larger portion, nearest the positively charged earth, developing a negative charge. The upper, broken off small droplets, separated from the tail and positively charged, are too small to continue falling. They are carried into the upper parts of the cloud. They may join other droplets, becoming larger, and fall back downward, becoming subject to the same charging and separation process.

As the negatively charged drops reach near the base of the cloud, they reach the zone of highest rising velocity, and may stall or stop falling in the rapid updraft. This negative charge accumulates at the cloud base.

The positive charges continue upward and diffuse throughout the upper portion of the cloud, thence drifting on up into an ionized layer many miles above the earth's surface. They return, broadly diffused in fair weather regions, to recharge the earth for its loss of positive charge from lightning.

The Earle R. Williams Article

"Although it has been known for two centuries that lightning is a form of electricity, the exact microphysical processes responsible for the charging of storm clouds remains in dispute."

The above quotation appeared as the subheading to an article entitled, "The Electrification of Thunderstorms," featured in the November 1988 edition of the *Scientific American* magazine. Written by geophysicist Dr. Earle R. Williams, a professor of meteorology at the Massachusetts Institute of Technology, the article is a lengthy discourse that goes deeply into charge distribution and configuration. In it, Dr. Williams discusses the basic structure of thunderclouds:

. . . years of observation have established that the basic structure of thunderclouds is not bipolar but tripolar: there is a main region of negative charge in the center with one region of positive charge above it, and a second, smaller region of positive charge below it. The most notable feature of the main, negatively charged layer is its pancake shape. Its vertical thickness is less than a kilometer, but it may extend horizontally several kilometers or more. It is in an altitude of approximately six kilometers, where the temperature is roughly –15 degrees Celsius. Under conditions prevailing there, all three phases of water—ice liquid and vapor—can coexist. The largest electric fields in the thundercloud are found at the upper and lower boundaries of the main negatively charged layer.

The upper region of the positive charge is more diffuse than the negative layer and may extend vertically several kilometers—as high as the cloud itself. The lower region of positive charge, on the other hand, is so small that the electric field at the surface of the earth is frequently dominated by the main negative charge.

It is estimated that nearly three billion lightning flashes bombard the earth in a year's time. Most are in tropical and temperate regions, few in cold areas. Because rain and lightning are both components of thunderstorms, wet regions have more lightning, and more lightning-caused destruction and casualties, than do dry areas.

Yet the latter do not escape the consequences of nature's most frequently occurring phenomenon. Even the sandy deserts of northern Africa have up to 10 thunderstorm days a year.

THE THUNDERSTORM DAY

A thunderstorm day (see Fig. 3.6) is a 24-hour period during which thunder is heard at a weather station serving as a counting center. The World Meteorological Organization and weather services of cooperating nations serve as a global recording society that keeps fairly accurate score on the number of days per year during which people in different localities hear the rumble and crash of thunder.

However, atmospheric scientists and weather experts are somewhat skeptical about the thunderstorm day's value. It is, they admit, the only monitoring system

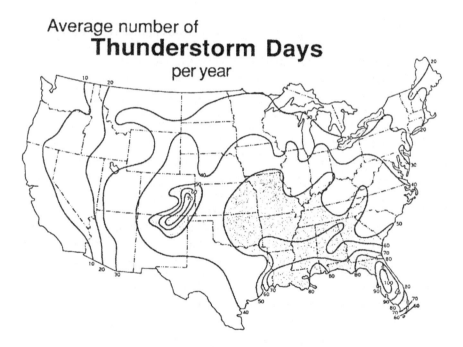

FIGURE 3.6 A thunderstorm day is a 24-hour period during which thunder is heard at a weather station serving as a counting center. This map showing storm frequencies in the United States is part of a worldwide recording program.

internationally available, but it does not give a very reliable lightning risk profile for any particular state or area. These are the chief criticisms:

- The thunder heard may be from either a cloud-to-cloud, intra-cloud, or cloud-to-earth discharge. Since the ratio of atmospheric as opposed to cloud-to-earth lightning flashes varies from region to region and season by season, the thunderstorm-day count produces incorrect comparisons of lightning activity in different localities.
- The count may be based on one crash of thunder at high noon or a hundred rumbles and crashes throughout the day. Upper New York state and the New England states, for example, have unusually intense and prolonged thunderstorms, and reliance on thunderstorm days as a sole guide is misleading in their particular cases.
- When using this calculation to establish risk, it should be borne in mind that the results are approximate because uniform terrain conditions are assumed. Higher and lower lightning flash rates may be observed as the result of micro-climates near mountains, near bodies of water, and near other nonuniform surface features.

Modern technology has provided systems that are improved over earlier calculation methods. Such systems triangulate on radio fields generated by a lightning strike and indicate both when and where the flash occurred. The radio frequencies of cloud-to-cloud and cloud-to-earth phenomena vary enough so that the new systems can differentiate between the two.

The thunderstorm-day count varies from one such day every 10 years at the North and South Polar regions to a recorded 242 thunderstorm days one violent year in Uganda, Africa (see Fig. 3.7).

The eardrums of townsfolk in Caruari, Brazil, are assaulted by crashes of thunder 206 days a year, while a Londoner and a resident of Beijing each can expect to hear thunder about 16 days per year. Paris, Rome, and Moscow hear the crash and rumble of atmospheric fireworks 21 days per year.

In the United States, Oakland and San Francisco, California, share a low of two thunderstorm days in the average year. In "Lightning Alley" across Florida, along a 50-mile wide swath from the Tampa area on the west to Daytona Beach on the east, thunder and lightning occur about 100 days per year. Fort Myers, south of "Lightning Alley" on the gulf side, is the lightning capital of the United States, with 102 thunderstorm days out of the annual 365 days.

While Florida residents hear thunder more often and suffer the greatest amount of lightning damage and casualties, it is midwesterners who undergo the most intense "squall-line" storm activity. Storms in Arizona, New Mexico, and Utah feature brisk lightning activity but little rain, which sometimes evaporates before it reaches the ground.

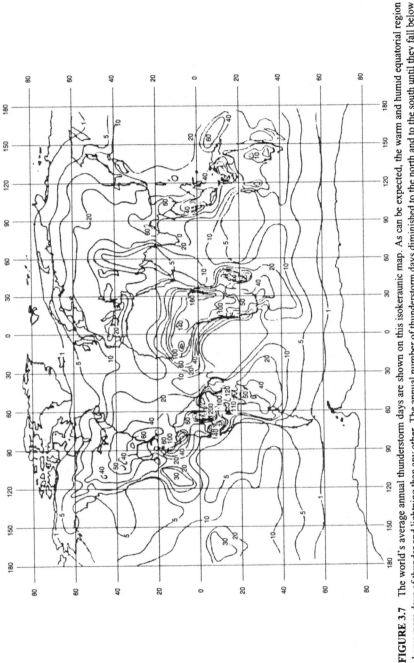

FIGURE 3.7 The world's average annual thunderstorm days are shown on this isokeraunic map. As can be expected, the warm and humid equatorial region shows more days of thunder and lightning than any other. The annual number of thunderstorm days diminished to the north and to the south until they fall below one day per year at the polar regions.

THE POINT-DISCHARGE PHENOMENON

During the days of sail, men on ships, traveling stormy seas under darkening skies, often saw what Christopher Columbus once described as a "ghostly flame which danced among our sails and later stayed like candlelights to burn brightly from the masts." Tending to be superstitious and at the same time being well aware of what lightning could do to wooden sailing ships, the sailors came to believe that the eerie glow was a sign of protection. They called it "St. Elmo's fire" after the patron saint of Mediterranean men of the sea. As it turned out, the sailors' faith was somewhat misplaced, because the "ghostly flame" consisted of point-discharge currents and sometimes foretold a cloud-to-ship lightning strike. Point-discharge currents are the usually positive earth's response to the compelling attraction of the negative charge at the base of a thundercloud above.

The earth, which is a huge electrical conductor, carries a slight negative charge in normal weather, while the upper atmosphere contains the opposite positive charge (see Fig. 3.8). The intermediate atmosphere, then, is a perpendicular positive field that, in fine weather, has a theoretically even electrical magnitude, variously quoted by atmospheric scientists as ranging from 100 to 600 volts per meter.

When a thundercell forms above, the earth soon loses its negative polarity and takes on a growing positive charge (see Fig. 3.9). The electric field between ground and thundercell is reversed, and the electrical potential between the two charge centers soars. Measured at a point just below the center of the negative cloud charge, the negative field from earth to cloud may reach a value as high as 10,000 to 20,000 volts per meter or more (see Fig. 3.10).

Visualize the electric field as consisting of evenly spaced horizontal lines. You can appreciate that a vertical object—a building, a flagpole, a person, or a lightning rod—will distort the electric field and cause the horizontal lines along the object's length to converge at its top (see Fig. 3.11). This produces an intense electric field at the top of the interrupting object. Obviously, the intensity will be related to the object's height and configuration. When the intensity of the electric field at the tip of the vertical object reaches a certain level, positive ions from the earth will stream upward, forming a point-discharge current or "corona" that will be released to the atmosphere.

Point-discharge currents are released by virtually all sharp objects on the ground, large and small. If they could be seen, the flows of current would resemble, as Columbus pointed out, "candlelights burning brightly from the masts." In fact, mountain climbers sometimes report seeing eerie point-discharge currents when conditions are right for the onset of a "mountain storm."

The positive currents tend to flow upward toward the attractive negative region at the base of the thundercell, and their tendency to do so grows as the strength of the electric field intensifies. If they could flow upward freely, perhaps the cloud-to-ground lightning phenomenon would not so often occur.

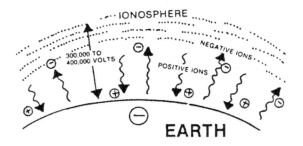

FIGURE 3.8 In normal weather, the earth carries a slight negative charge, while the ionosphere carries the opposite, positive charge.

FIGURE 3.9 Wherever thundercells exist, the ground below carries, for the moment, a positive charge of a magnitude determined by the amount of charge separation due to air turbulence.

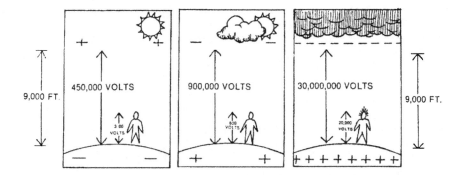

FIGURE 3.10 In normal sunny weather, the earth has a slight negative polarity, and the electrical potential difference between earth and atmosphere is about 50 volts per foot, as shown in the drawing at left. In the center figure, convection has caused air turbulence and a resulting switch in earth-sky polarities. At right is the electrical field as it might appear just before a lightning flash to earth of "typical" magnitude.

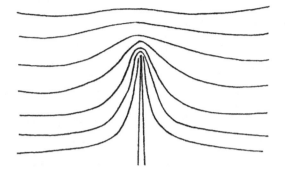

FIGURE 3.11   A vertical object like a tree, building, flagpole, person, or lightning air terminal will interrupt the electrical continuity of the air, bending the imaginary gradient lines of the electric field upward and causing an intense field concentration where the lines converge at the tip of the object.

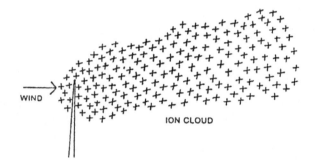

FIGURE 3.12   Positive ions are released into the atmosphere through development of point discharge currents as the electric field intensifies under a thundercloud. The ions want to stream toward their opposites in the cloud, but gusting winds carry the ion cloud off course as illustrated. The result is development of floating positive pockets in the air, attracting any downcoming negatively charged lightning flash and causing it to form the familiar zigzag pattern of descent.

WIND—THE KEY COMPONENT

But one of the components of the storm—in fact, the key component that triggers a thunderstorm in the first place—intervenes and prevents point-discharge currents from lessening the intense attraction between the center of the cloud mass and the positively charged ground directly below. That intervening component is wind—high, gusting wind. Carried off by gusts of wind, the released ions form pockets of positive space charges that float in the atmosphere (see Fig. 3.12). Meanwhile, back at the point where the discharged ions originated underneath the moving center of the negative atmospheric mass, the electrical intensity continues to increase, and the stage is set for eventual relief by a lightning flash.

4

The Lightning Flash

Lightning, which was believed by many pagan people to be the downward swoop of an avenging thunderbird, is still the subject of much confusion and incorrect assumption. For example, one rather widespread belief is that lightning will pursue ungrounded metal objects such as hand-carried golf clubs or radios.

Such notions may come about in understandable ways, however. Some people, including pro golfers Lee Trevino, Bobby Nichols, and Jerry Heard, have been injured by lightning-induced ground currents while holding golf clubs in contact with the ground at the instant of a nearby lightning discharge. Lightning-struck victims have exhibited burn marks where metal objects such as necklaces or watch chains have been in contact with the skin.

COMPONENTS OF A LIGHTNING FLASH

The transient discharge of electric current that occurs between a negative charge center and a positively charged region, commonly called a "lightning bolt," is in technical terms a *lightning flash*. A typical flash is made up of three or four discharge components or pulses, with each pulse consisting of a negative leader stroke and a positive return stroke.

The duration of a typical flash is 0.4 seconds. But this may vary from less than 0.1 seconds for a single pulse flash to a 0.5 seconds or more for a flash with up to 20 or more sequences of leader and return strokes. Some lightning flashes include negative strokes with low-magnitude continuing current that may stretch flash duration out to a full second or more. Such flashes are termed "hot bolts."

The rate of lightning flashes in thundercells of average activity is about two per minute. In severe storms, however, the flash rate may climb as high as 25 per minute, and in such storms the already severe hazard may be compounded by the fact that the more frequent flashes tend to be of longer duration as well.

A multistroke, cloud-to-ground flash may lower 25 coulombs of electric charge to ground per flash. Thus, 625 coulombs of charge may flow to earth per minute in a very severe thunderstorm with a flash frequency of 25 per minute.

Many lightning flashes appear to flicker. What is observed is a multiple stroke flash whose component strokes follow each other so swiftly that the eye is unable to discern anything other than a continuous but flickering spark.

In a typical lightning flash, the first leader stroke and subsequent leaders carry negative current downward. The first "stepped leader" stroke establishes the configurations of forks as well as the path of main current flow. The luminosity—the brilliant streak of light that you see—is caused by the return stroke, during which the negative current laid along the travel path flows to ground, and a positive "return stroke" flows upward.

The Stepped Leader Stroke

Atmospheric theory holds that a thundercloud primed to discharge has a tripolar configuration (see Fig. 4.1). It is generally agreed that a form of intercloud activity lasting up to 0.25 seconds precedes the initiation of the first stroke in a lightning flash. It is proposed that this prestroke activity consists of a breakdown of the air between the small positive charge at the lower leading edge of the cloud and the large negative charge center above it.

The resulting flash knocks electrons loose from the water drops, ice crystals, or snow particles to which they had become attached during the cloud separation process (described in the preceding chapter). The freed electrons overrun the small positive region, neutralizing it. Then, being at liberty to travel, the free electrons set off in an avalanche streamer toward the positive charge center on the ground below.

The downward trip of the stepped leader stroke is made in a series of discrete jumps varying in length from as little as 50 feet to as much as 900 feet, depending on electrical magnitude. The descent is usually erratic as the bright tipped front of the avalanching streamer pauses for (more or less) 60 microseconds (μs), fades out, then continues on with another bright tipped step, completing the descent at about one-thousandth the speed of light. Floating pockets of positive charge, per-

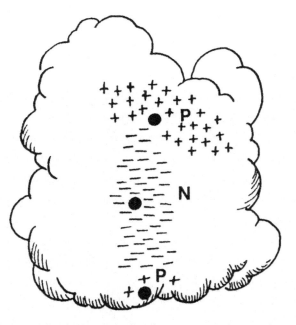

FIGURE 4.1   This is the electrical configuration of a thundercell that is primed for a lightning flash to earth. The upper cloud is positively charged, the lower portion is negatively charged, and the extreme lower edge carries a small positive charge.

haps previously emitted by sharp grounded objects, are neutralized along the stepped leader's path.

Some lightning flashes are highly branched. Others are not. Some meander widely. Others don't. The various branches continue toward earth and stop only when the main stream of electrons completes its trip. Apparently, flashes of high electrical intensity descend in relatively straight paths, while low-magnitude flashes sometimes appear to twist and turn, wandering hither and yon over large patches of sky (see Figs. 4.2 and 4.3). Typical cloud-to-ground travel time of a stepped leader is about 1 mile per 10 milliseconds, and the stepped leader distributes 2 to 3 coulombs of negative charge in the ionized channel over each mile of its length as it nears earth.

The Return Stroke

The large negative charge of the stepped leader induces a gathering of responding positive charges on the earth below, causing increasingly heavy point-discharge currents to flow upward on grounded objects. Positive streamers strain upward

38

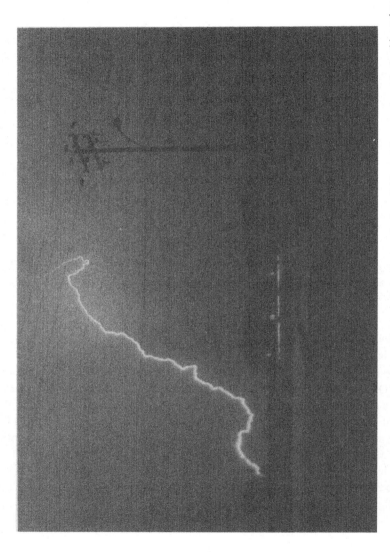

FIGURE 4.2 This meandering lightning flash may appear to be a single flow of current, but it probably contains several negative/positive pulses instead. Its torturous path suggests that it is a relatively weak flash whose stepped leader stroke is attracted hither and yon by positive charge areas in the atmosphere. Photo by David R. Stringer, *Daily Democrat*, Paul's Valley, Oklahoma.

FIGURE 4.3   The unbranched and relatively straight downward path of this flash indicates that it is a powerful one whose heavily charged stepped leader stroke headed directly for a correspondingly large concentration of positive charge directly below.   Photo by David R. Stringer, *Daily Democrat,* Paul's Valley, Oklahoma.

from the tips of trees, poles, the sharp corners and edges of houses, and other buildings, and even from animals and people (see Fig. 4.4).

Among this multiplicity of corona ribbons is one that, by location, elevation, and other factors of lesser influence (such as configuration and conductivity), shoots upward and succeeds in contacting the downward moving leader. Thus it determines the lightning strike point.

Immediately, a massive, intensely luminous return-stroke wavefront rises up the ionized path, traveling at a speed of from one-half to one-tenth the speed of light. This positive stroke makes the trip from ground to cloud in about 100 µs.

As the return stroke wavefront moves upward, heavy currents flow to ground along the entire length of that channel between ground and the moving wavefront. Thus the excess negative charge deposited along the length of the stepped leader stroke is brought to ground during the return stroke. Lesser currents may continue to flow for a brief period after the return stroke is completed (see Fig. 4.5).

Dart Leaders and Return Strokes

In a small percentage of lightning flashes, cessation of the current flow after the first return stroke may effectively drain the negative region in the thundercloud

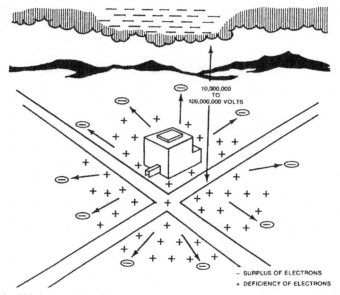

FIGURE 4.4   This is approximately what you would see if the electric field changes preceding a lightning flash were visible to the eye. As the negatively charged thundercloud passes above, it repels like charges and attracts positive charges.

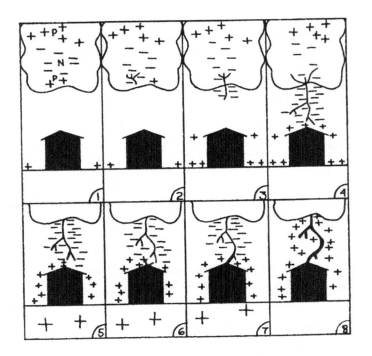

FIGURE 4.5   These drawings show (1) the typical charge configuration of a thundercell prior to a flash, (2) a local discharge between a small P region and the cloud's N region, (3) free electrons descending as they are attracted toward positive earth, (4) a stepped leader stroke that forks as it descends, (5) positive point-discharge streamers straining upward, (6) the descending negative stream as it nears effective ground, (7) a positive point-discharge streamer shooting upward, and (8) the return stroke as it makes the trip from earth to cloud at near the speed of light, as thousands of amperes of current flow down the channel to earth.

and complete the flash. But when additional negative charge is made available to the depleted region, at least one, and usually more, leader stroke and return stroke sequences will follow. It has been suggested that in cooler parts of the world, single-stroke flashes (that is, one stepped leader and one return stroke) are predominant, while multiple-stroke flashes are common in warmer regions.

When the cloud region that supplied the first stroke is left relatively uncharged in relation to some other portion of the cloud, the depleted region may be fed by a discharge between the top of the first stroke and a higher negative area. It has been noted that succeeding strokes in multiple-stroke flashes appear to drain higher and higher areas in the negative region of the cloud.

If a negative streamer from a higher area feeds electrons to a return stroke channel immediately, a leader stroke that is much faster than the stepped leader, and which travels downward continuously without steps, will occur. Called a *dart leader* because of its continuous downward travel, the second stroke will deposit a new negative charge that will flow to earth during the upward moving positive stroke. Usually, at least a third sequence of dart and return strokes will occur (see Fig. 4.6).

The return strokes zip up the ionized channel at about one-third the speed of light, leaving a trail of ionized air heated to 50,000° F, more or less.

Peak Currents

Current magnitude in the first return stroke will always be more severe than that of subsequent return strokes. It is theorized that half of all strokes have peak currents of 20,000 amperes (20 kA) or less. Ninety-nine percent of all strokes will have peak currents of 140 kA or less. Levels of 120 kA to 160 kA are routinely used when designing overhead shield systems.

For example, a target level of 99 percent reliability would be achieved by designing to a level of 140 kA, and 99.9 percent reliability would be obtained using a design level of 160 kA.

Stroboscopics and Other Effects

Frank C. Breeze, of the Florida based electrical engineering firm of Tilden, Lobnitz & Cooper, lectures on the subject of surge suppression in seminars held by the author, but he is also an expert on little-known effects of lightning:

> While driving toward a thunderstorm at night with windshield wipers set at their highest speed during rain, a stroboscopic effect will be achieved which causes your wiper blades to be seen in several positions at once. This result is due to lightning's multiple return strokes.
>
> Another lightning effect can be experienced by turning a radio on to a weak AM station or leaving it between stations. The crashing sound then heard between lightning strokes is due to the radiated electromagnetic field from the lightning channel, which induces a voltage on the car radio antenna.
>
> The crashing sound described is due to the fact that, while a lightning channel and an AM transmitting antenna are similar, magnitudes of current of the former are much higher, and the predominant frequencies are much lower. The radiated field strength from a lightning channel is also many orders of magnitude stronger than that of the largest broadcast transmitter.
>
> The majority of radiated energy falls at about 10,000 Hz (10 kHz) in the very low frequency (VLF) band. As the frequency increases, less energy is radiated. This explains why lightning is often seen as flashing noise in the picture of a low-band very high frequency (VHF) TV channel, which disappears when the set is tuned to higher channels.

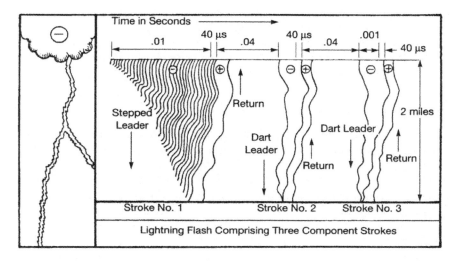

Time in Seconds ──────►

Lightning Flash Comprising Three Component Strokes

FIGURE 4.6   A fixed-lens camera sees a lightning flash as the human eye sees it. But a rotating-lens camera deflects the traces of a lightning flash sideways, revealing its component strokes, as represented in this drawing. The invention of Sir Charles Boys, the camera has been used since 1933 to help unveil the secrets of lightning.

Field Intensity Changes

Still quoting Mr. Breeze,

The field intensity of lightning current changes appreciably with distance. To put this into perspective, suppose lightning were to strike about 100 meters away from a building. Statistical data is obtained that shows a field intensity of 1 volt per meter at a distance of 10 kilometers. The distance from the building in question, however, is 100 times closer. The increase in amplitude, therefore, will be 100^2, or 10,000 volts per meter. A directly buried cable or one encased in plastic conduit will develop significant amounts of voltage along its length when located in such a strong, changing field. These voltages will be presented to equipment at both ends of the circuit.

Another very important characteristic of a lightning flash is the waveform of its return stroke. Figure 4.6 shows the rise and decay of current involved in a typical return stroke. The rapid rise of current from zero to crest value makes lightning current very difficult to handle in grounding and bonding circuits.

The waveforms shown in Fig. 4.6 have risetimes averaging 1.5 microseconds. Risetime is described as the time it takes for current amplitude to increase to 90 percent of its maximum value. Decay time is measured to the point where amplitude diminishes to 50 percent of its crest value. In the case illustrated, decay of the average waveform is shown as about 8 microseconds. This type of double exponential waveform will often be referred to as a 1.5 by 50 microsecond waveform, 8 by 20 microsecond waveform, or other description using two such numbers. The first number always represents the risetime, and the second number represents the decay time.

The waveform shown, while accurate for an actual lightning stroke, will differ once the lightning current enters the sphere of circuits. Inductive and capacitive properties of these circuits will cause the waveform to vary.

Strokes With Long Current Tails

"Hot bolts" of lightning are atypical flashes that include strokes that lower negative charges of about 30 coulombs of charge to ground, at a rate of perhaps 900 amperes (A) of current, over periods in the vicinity of 0.2 seconds. These long-duration strokes are the primary causes of forest fires. They also ignite thousands of buildings each thunderstorm season. The heating effect of the relatively low-amperage current flowing for long intervals is cumulative. Scientists estimate that low-amperage current strokes occur from 20 to 35 percent of the time in multiple-stroke flashes.

Flashes with Multiple Strike Points

Lightning sometimes will strike more than one point. Photographs have shown instances in which current flow branching has taken place near ground level. In one case, a chance snapshot showed a branched flash killing one person and severely injuring another standing at least 15 feet away.

There have been a number of instances where lightning has struck and killed or severely injured one person while felling others in the immediate vicinity. Such instances are not rare, and they occur most often at football, baseball, or other sports-related locations. The principal victim is struck directly, and the others are victims of upward-darting currents leaving their heads to join the main return-stroke current flow. Almost always, the secondary victims suffer headaches, and sometimes a term of disorientation, from the effects. These currents that may leave them unconscious for a time, with lingering amnesia.

Recently, scientists have concluded that, in some instances, succeeding strokes will follow a different path to ground from the one established by the first sequence of stepped leader and return strokes. The reason given is that the lower end of the channel cools so rapidly that ionization decays, and the path a dart leader sets out to follow effectively disappears. At some point, then, the dart leader becomes a new stepped leader, and in a series of discrete downward leaps, establishes a new ionized path that its pulse companion, a new return stroke, will follow.

The Triggered Upward Flash

To anyone favoring lightning as a photographic subject, the dramatic and relatively rare shot of a flash with upward-reaching branches is a prime trophy. Luck, as well as a lot of patience, is required, because it is estimated that such flashes represent only 5 to 10 percent of all discharges between clouds and earth.

The likeliest spot to capture such a lightning flash with a camera is a vantage point with a clear view of a very tall building or transmitting tower. By their very heights such structures actually trigger lightning discharges (see Fig. 4.7).

Atmospheric scientists have found that for structures 300 feet high or less, triggered lightning strikes are a negligible factor compared to discharges initiated by downward-traveling stepped leaders. But at heights of 700 feet or more, structures receive more triggered lightning flashes than cloud-initiated flashes. On mountain peaks, which sometimes jut into or extend above thunderclouds, even a flagpole or ordinary building can trigger an upward-moving flash.

A typical triggered flash begins with a heavily branched upward-moving leader stroke, which may be either negative or positive. In many cases, there is no return stroke; instead, the upward-moving leader's luminous channel fades away.

In other instances, the upward stroke is followed by one or more downward, negative dart leader strokes, each of which is followed by a positive return stroke. The result is deposition of excess negative charge along the channel by each leader stroke and the release of the charge to ground during each return stroke.

In effect, then, an upward-moving triggered lightning flash poses the same potential danger to people and property as a "normal" cloud-initiated flash.

Lightning's Electrical Dimensions

While lightning currents are difficult to measure, it is generally agreed that about 20,000 A is typical, and that flows of more than 80,000 A are rare. The maximum value of any discharge computed thus far by a reliable source is 340,000 A. Many lightning flashes occur that transmit weak currents, on the order of 1,000 to 3,000 A, and those are not included in establishing what is "typical."

A critical dimension of lightning is the rate of rise to peak current flow during the return stroke. Scientists had assumed that the current wavefront reaches its crest in 1 to 3 μs, But recently, the suggestion has been made that the time element is less than 1 microsecond.

In multiple-stroke flashes, the wavefront of the current flow during the first return stroke has been shown to have a concave shape, increasing in steepness as it rises to its maximum value in from 15 to 20 μs, then holding at that current rate, or near it, for perhaps 20 μs. It then falls off to about 20 percent of its crest value in 200 to 300 μs. Subsequent stroke currents have much faster rise times, reaching crests in from 1 to 2 μs and falling off more quickly.

Another critical characteristic, which is dependent on the amount and duration of current flow, is temperature. Scientists, using various methods to arrive at their conclusions, have come to general agreement that temperatures up to 30,000° F are typical, but that they may rise to higher readings, even 50,000° F.

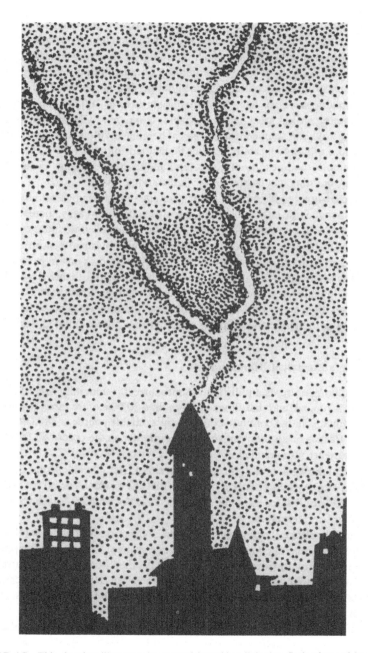

FIGURE 4.7   This drawing illustrates an upward branching lightning flash triggered by a tall building.

Striking Distance

The *striking distance* of lightning is the distance between the tip of the downcoming stepped leader and the point to be struck (see Fig. 4.8). The striking distance is roughly equal to a step's distance which, as the leader nears ground, may be somewhat greater than the length of the average step of that stroke.

Striking distances increase with increased current amplitude and range from as little as 16 m for a stroke of about 10,000 A of negative current to more than 200 meters for a 160,000 A stroke of positive polarity.

Lightning Frequency

A typical thunderstorm cell has a life span of 30 to 60 minutes, with a flash rate of 2 to 3 per minute. A storm may consist of several active cells at any time, and the total number of flashes therefore depends on the number of active cells and the storm's duration.

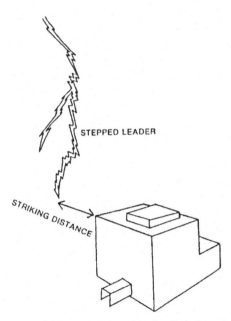

STEPPED LEADER

STRIKING DISTANCE

FIGURE 4.8   The striking distance of a particular lightning flash depends on the amplitude of the flash. The stepped leader stroke of a flash of "typical" amplitude descends in steps about 150 feet in length. Therefore, 150 feet is viewed as the minimum distance of attraction between a negative stepped leader and an earthbound, positively charged object.

A two-cell convective storm with a one-hour duration will contain between 480 and 720 lightning flashes, and if the ratio of cloud–ground to cloud–cloud flashes is the typical U.S. ratio of 1 in 3 or 4, there will be from 120 to 240 flashes to ground during those 60 minutes.

However, no two storms are alike. A very severe storm in a temperate region may have a flash rate as high as 10, 15, or even 25 per minute. And since the ratio of cloud–ground to cloud–cloud discharges is higher in temperate zones, hundreds of flashes may strike to ground in an hour's time.

Central Florida has nearly 100 thunderstorm days annually, while Massachusetts has 20 days annually (i.e., days when thunder is heard at a weather station in the state). However, Florida has a lower than average ratio of cloud-to-ground flashes, while northern Atlantic coastal regions have a higher than average ratio. Florida's ground-strike expectations, as computed by the world average, are reduced, while Massachusetts' ground-strike expectations are raised.

Strikes to Tall Structures

Tall structures that trigger upward-moving lightning leaders can expect many more lightning strikes than lower buildings, because they will receive their share of discharges due to their "attractive radii" as well as a certain number of triggered flashes, dependent on height.

Atmospheric physicist Dr. Rodney Bent used as an example a 1,200-foot tower at Orlando, Florida, where the incidence of lightning flashes to ground per year is given as 7.5 per square kilometer. Due to its height of 365 m, the tower would have an attractive radius of some 400 m and therefore could expect to be the target of 3.77 cloud-initiated strikes per year. In addition, the tower could be expected to trigger about 39 upward-moving flashes annually, for a total per year of 43.

The Sound of Lightning

"Any of the dry exhalations that get trapped when the air is in the process of cooling is forcibly ejected as the clouds condense and in its course strikes the surrounding clouds, and the noise caused by the impact is what we call thunder." So wrote the great ancient philosopher Aristotle during the fourth century B.C. But in 1888, the *Scientific American* carried an article giving this more accurate explanation, by M. Hirn:

. . . . The sound which is known as thunder is due simply to the fact that the air traversed by an electric spark, that is, a flash of lightning, is suddenly raised to a very high temperature, and has its volume, moreover, considerably increased. The column of gas thus suddenly heated and expanded is sometimes several miles long, and as the duration of the flash is not even a millionth of a second, it follows that the noise bursts forth at once from the whole column, though for the observer in any one place it commences where the lightning is at the least distance.

The beginning of the thunder clap gives us the minimum distance of the lightning, and the length of the thunder clap gives us the length of the column.

So it is. The loud crash of a nearby lightning strike is heard at the same time as the flash is seen, but if the flash channel is 2 miles long, you will hear the rumble of that nearby strike for another 2 seconds.

Lightning travels at a speed of 186,000 miles per second, while sound moves along at a comparative snail's pace of 1,090 feet per second, or about one mile per 5 seconds. Therefore, you can use your watch to calculate roughly how far away a lightning flash is, counting 5 seconds per mile (see Fig. 4.9).

Superbolts, Ball Lightning and Other Phenomena

Every once in a great while, a lightning flash will occur that begins with a stepped leader carrying a positive charge to earth. Originating from the positive charge re-

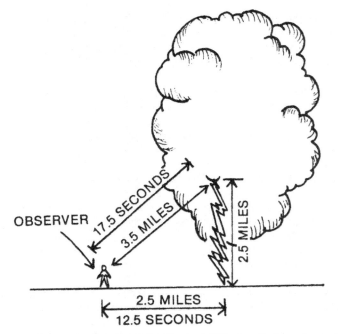

FIGURE 4.9   How far away is a lightning flash? About one-fifth of a mile for each second it takes the sound of thunder to reach you. Why does distant thunder "rumble," while nearby thunder "crashes?" Because you hear the sound of exploding air at the base of the column several seconds before the sound of the more distant top of the lightning column reaches you. This contrasts to nearby thunder, where the crash of the entire column reaches you almost immediately.

gion of the cloud, the flash is positively charged by virtue of a negative charge flow out of the top of the leader into its positively charged point of origin.

Such flashes are "superbolts" that transfer some three times as much charge as do negative strokes. They have long wavefronts, due to their slow rates of rise and long wave tails, transferring their large charges over extended stroke durations. It is believed that positive charges rarely consist of more than one stroke.

The much discussed phenomenon known as "ball lightning" has been seen so often by people around the world that science has come to take it seriously, although thus far science has not agreed on any of the explanations offered for its existence (see Fig. 4.10). Typically, ball lightning is described as about the size of a grapefruit. It usually is seen near the ground, moving horizontally and exhibiting hissing sounds, heat, and odor. It is reported as having a lifetime of a few seconds, and it either silently fades away or seems to explode.

"In Search of Fireball Lightning"

Robert K. Golka, with Dr. Robert W. Bass, headed *Project Tesla,* a 1970s attempt to artificially duplicate ball lightning. Golka has been indefatigable in his quest and, finally, he was successful during a remarkable ride in a diesel locomotive. The following excerpts from his article in the March 1985 edition of *Radio-electronics* magazine describe his pioneering journey.

About two years ago, while going through Dr. Tuck's personal papers (Dr. James L. Tuck, of World War II Manhattan Project fame and the founder of the Plasma Physics Laboratories at Los Alamos National Laboratories), I came across four sequential photos of a submarine battery-bank circuit breaker being opened; the current through the breaker was on the order of 10,000 amps. Dr. Tuck had shown those pictures to me in 1971, and mentioned that after about 30,000 shots over a $2^1/2$-year period, he had gotten only those 4 pictures of what he thought to be ball lightning. . . .

Upon seeing those photographs once again, I decided to reproduce Dr. Tuck's experiments using a real World War II submarine. Unfortunately, there are few of those left in our 'nuclear Navy'. The only one I managed to locate was not being used, but that may have been due to the fact that it had no propellers and only one working engine. I thus decided to try the experiment with a diesel railroad locomotive instead.

The next problem was finding a railroad that was willing to loan me a locomotive, box cars, and track for such an experiment. Thankfully, the president of the Boston and Maine railroad had the wisdom, the foresight and love of progress necessary to fulfill my needs; the railroad provided me with two locomotive, a train, and a mile and a half of track. It was because of that, that I was able to conduct $3^1/2$ months of experiments that completely altered my thinking on the nature of ball lightning.

First of all, I now feel that it is more of a particle rotation flow than a high voltage electrostatic effect; that is, more like a giant plasma vortex donut with a tiny hole than an electrostatic sphere. Now, there are a whole host of phenomena in aeronautical engineering, particularly in the area of fluid dynamics, that are not yet fully understood.

FIGURE 4.10 Is "ball lightning" an actual phenomenon, or is it merely a lingering image on the retina of the viewer's eye? Scientists have come to at least suspect that the elusive phenomenon exists and someday may be explained.

One of those is the physical properties of vortices. One can blow smoke rings inside of smoke rings, and have the inner ring move back and forth. You can blow smoke rings that stand perfectly still. In liquids, rings can form spheres and other shapes.

Well, back to the railroad! To perform my experiments, I grafted a submarine circuit breaker into the high-voltage circuit between the million-watt, 1600-horsepower diesel generator and 2000-horsepower motor trucks beneath the locomotive. By opening the circuit breaker (using a long broomstick handle), I was able to generate ball lightning.

The effects of opening the circuit breaker were quite astonishing. Temperatures in the cab of the locomotive would go instantly from 60° F to 110° F. As you might imagine, there was an overwhelming desire to leave the train cab for some fresh air. I, of course, could not do that since the train was still moving (at about 20 miles an hour), and the likely result would have been running the train off the end of the track and destroying the experimental setup.

In any event, after redoing the experiment countless times, I was able to convince myself that the fireball effect was due to the elimination of turbulence. In effect, when I closed the door and windows of the cab, the effect was most likely to occur.

My earlier experiments involved placing the breaker on the top of the locomotive, over the updraft of the engine radiators; that proved to be the most ineffective way to go. It was only after I realized that shielding the setup from turbulence aided the effect that I began to see results. If nothing else, that shows that one must have almost a Sherlock Holmes-like approach to this kind of research if it is to succeed. Also, that was probably the first plasma physics experiment ever performed on a moving train!

I now feel that there is much to be done in this area. The next step is to perform the experiment using a more conventional setup: in a low velocity wind tunnel, using controlled arc discharges to form ball lightning. I believe that setup would be very productive.

While the mechanism that allows the confinement of the plasma is still unknown, now the probable nature of that mechanism is known, at least in part. The next task is to demonstrate the phenomenon over and over again. By doing that, the remaining questions will be answered and the true nature of ball lightning will be revealed."

The author's wife once had a vividly remembered experience with ball lightning. While sitting near the kitchen door, she saw and heard such a ball, which she describes as "like a glowing ball of fire" that came in from outside the doorway and rolled past her along the floor to the drain pipe below the sink, where it disintegrated. It left her with a tingling feeling in one hand, with which she was holding a pair of scissors.

Other Lightning Phenomena

Several other lightning phenomena have been observed.

- *Heat lightning*. This is explained as merely cloud illumination due to reflection from a lightning flash which is too distant for thunder to be heard.
- *Sheet lightning*. This, too, is described as merely cloud illumination due to far-off atmospheric discharges.
- *Ribbon lightning*. This is a multiple-stroke flash whose component strokes shift horizontally, apparently due to wind. The human eye sees the separate strokes individually but simultaneously, not distinguishing the time intervals separating them.
- *Bead lightning*. Bead lightning is a flash whose channel appears to break up into fragments of about one step in length. Several theories have been offered in an effort to explain the phenomenon. One holds that the eye sees parts of the channel from the side and others head on, due to the twisting and turning typical of a stepped-leader stroke. Another explanation is that clouds or rain mask parts of the channel. Still another explanation is that some of the channel's steps remain luminous longer than others. Bead lightning is typical of thunderstorm phenomena: it comes and goes often enough to become a rather familiar mystery, but not often enough to explain itself, even through what has become a remarkable discipline—atmospheric science.

Flash Rates in U.S. Cities

The data in Table 1 are offered for use in estimating the number of lightning related problems to be expected in the vicinities of various cities in the United States. Shown are flash rates for two cities in each state, with a few exceptions.

The cities involved are those with the greatest number of flashes and the least number of flashes annually.

The *isokeraunic level* refers to the average number of times thunder is heard in the particular city's environs during a year's time. Figures in the *total flashes per year* and the *flashes to ground per year* columns are also based on averages. The data has been derived from statistics gathered by the World Meteorological Organization (WMO) ove· many years. As such, it is average data and the experience in any one year may differ somewhat from that predicted.

TABLE 4.1   World Meteorological Organization Lightning Statistics

City	Isokeraunic Level	North Latitude	Total Flashes per Year		Flashes to Ground per Year	
			per km²	per mi²	per km²	per mi²
Alabama						
Birmingham	67	33°40'	25.4	65.9	5.7	14.8
Montgomery	54	32°18'	17.6	45.6	3.8	9.9
Arizona						
Prescott	43	34°39'	12.0	31.0	2.8	7.2
Yuma	10	32°45'	1.0	2.6	0.23	0.6
Arkansas						
Fort Smith	53	35°22'	17.1	44.2	4.2	11.0
Texarkana	71	35°00'	28.1	72.7	6.2	16.1
California						
Mt. Shasta	14	41°17'	1.8	4.6	0.5	1.3
S.F./Oakland	2	37°45'	0.1	0.2	0.02	0.04
Colorado						
Colorado Spr.	68	38°49'	26.1	67.5	7.0	18.2
Grand Junction	41	39°06'	11.0	28.6	3.0	7.7
Connecticut						
Hartford	27	41°44'	5.4	14.0	1.6	4.1
New Haven	24	41°16'	4.4	11.5	1.3	3.3
Delaware						
Wilmington	33	39°48'	7.6	19.8	2.1	5.4
D.C.						
Washington	35	38°51'	8.4	21.8	2.3	5.9
Florida						
Daytona Beach	93	29°20'	44.4	115.0	8.7	22.5
Miami	70	25°49'	27.4	71.0	4.8	12.4
Pensacola	70	30°21'	27.4	71.0	5.5	14.3
Georgia						
Augusta	41	33°28'	11.0	28.6	2.5	6.4
Valdosta	69	30°53'	26.7	69.2	5.5	14.3

TABLE 4.1   World Meteorological Organization Lightning Statistics (continued)

City	Isokeraunic Level	North Latitude	Total Flashes per Year		Flashes to Ground per Year	
			per km²	per mi²	per km²	per mi²
Idaho						
Lewiston	17	45°58'	2.5	6.4	0.8	2.1
Pocatello	27	42°55'	5.4	14.0	1.7	4.3
Illinois						
Cairo	58	37°00'	19.9	51.5	5.0	13.0
Chicago	37	41°47'	9.3	24.0	2.7	7.1
Indiana						
Evansville	50	38°02'	15.5	40.0	4.0	10.4
Ft. Wayne	41	41°10'	11.0	28.6	3.7	8.2
Iowa						
Burlington	56	40°47'	18.7	48.6	5.3	13.8
Dubuque	39	42°24'	10.1	26.3	3.1	7.9
Kansas						
Dodge City	39	37°46'	10.1	26.3	2.6	6.8
Wichita	54	37°38'	17.6	45.6	0.5	11.7
Kentucky						
Lexington	44	38°02'	12.4	32.2	3.2	8.4
Louisville	46	38°11'	13.4	34.8	3.5	9.1
Louisiana						
Baton Rouge	78	30°25'	32.9	85.3	6.7	17.3
Lake Charles	78	30°13'	32.9	85.3	6.6	17.1
Shreveport	50	32°33'	15.5	40.0	3.4	8.7
Maine						
Eastport	13	44°54'	1.6	4.1	0.5	1.3
Portland	27	43°39'	5.4	14.0	1.7	4.4
Maryland						
Baltimore	32	39°11'	7.2	18.8	2.0	5.1
Frederick	24	39°20'	4.4	11.5	1.2	3.1
Massachusetts						
Pittsfield	29	42°25'	6.1	15.9	1.9	4.8
Salem	5	42°28'	0.3	0.8	0.1	0.2
Michigan						
Alpena	24	45°04'	4.4	11.5	1.4	3.7
Lansing	40	42°47'	10.6	27.4	3.2	8.3
Sault Ste. Marie	24	46°25'	4.4	11.5	1.5	3.9
Minnesota						
Int'l. Falls	28	48°36'	5.8	14.9	2.1	5.4
Rochester	40	44°00'	11.0	27.4	3.3	8.6

TABLE 4.1   World Meteorological Organization Lightning Statistics (continued)

City	Isokeraunic Level	North Latitude	Total Flashes per Year per km²	Total Flashes per Year per mi²	Flashes to Ground per Year per km²	Flashes to Ground per Year per mi²
Mississippi						
Jackson	64	32°20'	23.5	60.9	5.1	13.2
Meridian	64	32°20'	23.5	60.9	5.1	13.2
Vicksburg	62	32°24'	22.3	57.7	4.8	12.5
Missouri						
Springfield	59	37°14'	20.5	53.1	5.2	13.5
St. Louis	49	38°45'	14.9	38.7	4.0	10.3
Montana						
Butte	43	46°00'	12.0	31.0	4.0	10.4
Kalispell	22	48°11'	3.8	9.9	1.4	3.5
Nebraska						
Norfolk	53	41°59'	17.0	44.2	5.1	13.1
North Platte	38	41°08'	9.7	25.1	2.8	7.2
Nevada						
Ely	31	39°17'	6.9	17.8	1.9	4.9
Winnemucca	11	40°54'	1.2	3.1	0.03	0.9
New Hampshire						
Mt. Washington	16	44°16'	2.2	5.8	0.7	1.8
New Jersey						
Atlantic City	23	33°22'	4.1	10.7	1.1	2.9
Trenton	35	40°13'	8.4	21.8	2.4	6.1
New Mexico						
Raton	75	36°58'	30.8	79.8	7.8	20.1
Roswell	45	33°24'	12.9	33.5	2.9	7.5
New York						
Albany	23	42°45'	4.1	10.7	5.4	3.2
Binghamton	31	42°05'	6.9	17.8	2.0	5.3
N.Y.C.	31	40°46'	6.9	17.8	2.0	5.1
N. Carolina						
Asheville	53	36°36'	17.1	44.2	4.1	10.6
Cape Hatteras	40	35°15'	10.6	27.4	2.5	6.5
N. Dakota						
Bismark	31	46°46'	6.9	17.8	2.4	6.1
Williston	25	48°09'	4.8	12.3	1.7	4.4
Ohio						
Cincinnati	53	39°04'	17.1	44.2	4.6	11.9
Sandusky	31	41°25'	6.9	17.8	2.0	5.2
Oklahoma						
Okl. City	45	35°24'	12.9	33.5	3.1	8.0
Tulsa	58	36°11'	19.9	51.5	4.9	12.6

TABLE 4.1   World Meteorological Organization Lightning Statistics (continued)

City	Isokeraunic Level	North Latitude	Total Flashes per Year		Flashes to Ground per Year	
			per km²	per mi²	per km²	per mi²
Oregon						
Baker	16	44°50'	2.2	5.8	0.7	1.9
Eugene	5	44°07'	0.3	0.8	0.1	0.3
Roseburg	5	42°13'	0.3	0.8	0.1	0.2
Pennsylvania						
Curwensville	47	40°59'	13.9	36.0	4.0	10.3
Williamsport	20	41°15'	3.3	8.4	0.9	2.4
Rhode Island						
Block Island	17	41°10'	2.5	6.4	0.7	1.8
Providence	21	41°44'	3.5	9.2	1.0	2.7
S. Carolina						
Charleston	56	32°54'	18.7	48.6	4.13	10.7
Columbia	47	33°57'	13.9	36.0	3.2	8.3
Florence	56	34°11'	18.7	48.6	4.3	11.1
S. Dakota						
Huron	38	42°23'	9.7	25.1	3.1	8.0
Rapid City	41	44°09'	11.0	28.6	3.5	9.0
Tennessee						
Chattanooga	58	35°02'	19.9	51.5	4.7	12.2
Knoxville	48	35°49'	14.4	37.4	3.5	9.0
Texas						
Del Rio	27	29°20'	5.4	14.0	1.0	2.7
Port Arthur	72	29°58'	28.7	74.4	5.8	14.9
Utah						
Milford	28	38°24'	5.8	17.9	1.5	3.9
Salt Lake City	35	40°46'	8.4	21.8	2.4	6.2
Vermont						
Burlington	28	44°28'	5.8	14.9	1.9	4.8
Virginia						
Lynchburg	35	37°20'	8.4	21.8	2.2	5.6
Roanoke	42	37°19'	11.5	29.8	2.9	7.6
Washington						
Ellensburg	11	47°02'	1.2	3.1	0.4	1.1
Olympia	3	47°00'	0.1	0.3	0.04	0.1
Spokane	11	47°33'	1.2	3.1	0.4	1.1
Tatoosh Island	3	48°23'	0.1	0.3	0.03	0.1
W. Virginia						
Charleston	47	38°22'	13.9	36.0	3.7	9.5
Parkersburg	43	39°21'	12.0	31.0	3.2	8.4

TABLE 4.1   World Meteorological Organization Lightning Statistics (continued)

City	Isokeraunic Level	North Latitude	Total Flashes per Year		Flashes to Ground per Year	
			per km²	per mi²	per km²	per mi²
Wisconsin						
Green Bay	32	44°29'	7.2	18.8	2.3	6.0
Madison	43	43°0.8'	11.0	28.6	3.4	8.8
Wyoming						
Cheyenne	46	41°09'	13.4	34.8	3.9	10.0
Lander	22	42°48'	3.8	10.0	1.2	3.0

5

Lightning Injuries, Fatalities, and Public Safety

Among certain pagan people, lightning was long accepted as a form of swift and certain punishment by wronged and wrathful gods. Survivors of lightning strikes were shunned as unclean, while the apparently dead were sometimes buried on the spot without benefit of pomp or ceremony.

We now know, based on experience and the teachings of first aid, that lightning victims who exhibit signs of life should be left to struggle toward dazed and tottering recovery by themselves, as the apparently dead are given cardiopulmonary resuscitation (CPR) to restore breathing and/or heart action.

Here in America, reaction to the threat of lightning death or injury is mixed. On the one hand, many people still dash to take cover under a lone, conspicuous, unprotected tree to escape pelting rain. Each year, scores of Americans pay a fatal price for such ignorance of lightning safety rules.

On the other hand, American lawyers, in several recent cases, have broken down the "act of God" defense that, since the advent of English tort law, has held blameless the owners of unprotected shelters that have been the sites of lightning casualties. The spectre of liability, then, is quite real when people are exposed to possible lightning injury or fatality on another's unprotected property.

THE DANGEROUS VOLTAGES

A person crossing an open field while riding a motorcycle; a horse; or an industrial, farm, or recreational vehicle without a metal cab or roof may be the most conspicuous target in the immediate area. When lightning current flashes down in the vicinity, he may be subject to a direct lightning strike. However, the majority of lightning deaths and injuries are not due to direct strikes but to the indirect effects of flashes to other and more prominent targets, and to the secondary effect of lightning-induced fire.

Lightning current can cause serious burns and damage to the heart, lungs, central nervous system, and other body parts through heating and a variety of electrochemical reactions. How much injury will be sustained depends on the amount of current flow, body parts affected, the physical condition of the victim, and the particular circumstances of the incident (see Fig. 5.1).

Lightning's perils consist of dangerous voltages that can cause problems through (1) direct strikes, (2) secondary effects, and (3) indirect contact.

Direct Strikes

A person can be the target of a direct strike under several different conditions, in addition to those cited above. A fisherman in an open boat, a mountain climber, a person jogging on an open road, a cyclist, a horse rider, and work personnel atop roofs or utility poles expose themselves to the dangers of direct lightning current when they do not take proper shelter during lightning storms (see Figs. 5.2 through 5.6).

Secondary Effects

By far the most common secondary effect of lightning current is death or injury due to burns or asphyxiation in lightning-induced fires. Less common examples are electrocution from lightning downed power lines and auto accidents due to temporary blindness when lightning strikes a speeding automobile or another object in the vicinity (see Fig. 5.7). In one such instance, lightning struck in front of an automobile, temporarily blinding the driver and causing six injuries in the ensuing accident.

In another incident, lightning struck a tree, which toppled over a high-voltage power line, which snapped and curled over an automobile, electrocuting the driver. Another incident occurred on a winding New England road. Lightning struck a tree, which fell across the highway atop a passing automobile, crushing and killing the driver. His wife, sitting alongside, survived.

FIGURE 5.1   This shirt was worn by the Reverend Jerry H. Atkins when he was struck by lightning at Tanglewood Park, Winston-Salem, North Carolina. He had heard thunder some distance away and had sat down on a bench when struck. He was paralyzed for two hours, was unconscious briefly, and suffered skin burns as the current coursed along his skin.   Photo by Winston-Salem, North Carolina, *Journal*.

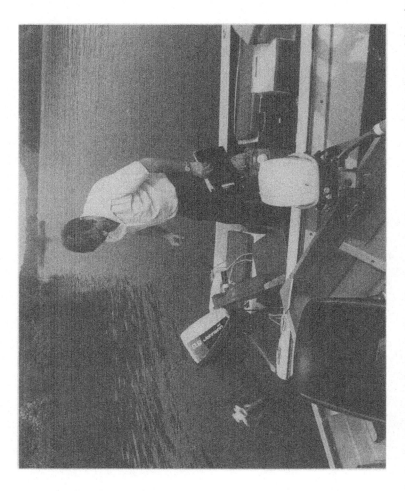

FIGURE 5.2 Butler County, Ohio, coroner William F. Young, Jr., checks the boat in which a man was killed as he operated the outboard motor. Two boys riding with him were unhurt. Two factors played parts in the tragedy: the man's larger size as a target, and the churning of the water. Photo by Butler, Pennsylvania, *Eagle*.

FIGURE 5.3   Two people boating on the canal in the background sought cover under this tree. Both suffered fatal sideflashes of current when the tree was struck by lightning.

FIGURE 5.4   More injuries occur in lightning-ignited houses than in any other location. A major reason is that lightning often starts electrical fires that smolder inside the walls until they eat through and erupt into furious blazes.   Photo by Gina Wright, Louisa, *Kentucky, Big Sandy News.*

Indirect Contact

Fatal or injurious currents can be sustained through indirect contact under several different conditions, as described below.

Induced Voltage

The strong electric field produced at ground level as a leader stroke descends toward a strike point grows increasingly intense and may induce sufficient current in a nearby person's body to cause a positive streamer to flow upward from his head. Death sometimes results, particularly in the case of an individual who is aged or in poor health.

It is not uncommon for a number of persons to suffer the effects of upwardly flowing ground currents as lightning is striking nearby. Many instances have occurred where an individual has been struck directly and killed while several people in the group have suffered painful headaches and perhaps a period of amnesia. In such instances, the survivors suffer the effects of induced voltage streamers flowing across their bodies and their heads to join the main upward streamer. Typically, such streamers flow at about 1,000 volts (V) of potential.

FIGURE 5.5   The empty chain hooked to this doghouse is mute testimony to lightning's ability to kill through ground voltage. The dog secured by the chain died when lightning struck the Betty Jo and Ernest Hudson house for the third straight year.   Photo by Joy M. Hall, Lucedale, Missouri.

Flashover Voltage

People can become conductors of direct current when (1) lightning strikes a tree and takes a parallel path through the body of someone standing or sitting in close proximity to the tree's trunk; (2) lightning current flashes over to a person standing or sitting inside a small, unprotected building like a golf course shelter; or (3) current sideflashes to a person from a conductor like a plumbing line or metal kitchen sink, a metal vent, or a carbon-laden fireplace flue; or when current sideflashes to a person immediately adjacent who is struck directly by lightning (see Fig. 5.8).

Among the legal cases involving lightning-caused casualties where the author served as an expert witness have been incidents of injury due to sideflashes. In one particularly tragic instance, a young couple with minor children were killed when lightning struck a small tree about 3 feet from the wall of a golf course shelter where they had taken refuge from rain. Lightning current flashed to their heads through the open space above a low wall.

In another incident, a parishioner had stepped outside for fresh air during a church service. Lightning struck a new fiberglass steeple atop the roof and travelled down the conductor that steeple manufacturers often build into such towers.

FIGURE 5.6   Lightning during a nighttime thunderstorm killed these valuable horses in a pasture near West Plains, Missouri. Stable owner Frank L. Martin could not determine the lightning strike point.   Photo courtesy of *Daily Quill,* West Plains, Missouri.

FIGURE 5.7 Cheryl Bircher, of Green River, Wyoming, was driving this pickup truck when she heard a crash "like a big explosion." She thought at first that another car had rammed into her truck until she saw the truck's burned-off antenna. The antenna was melted to the base, and the vent window and sideview mirror were shattered. Photo by Ken Smith, *Green River Star*.

FIGURE 5.8   A fairly common cause of lightning shock or injury in an unprotected or improperly protected house is exemplified by this sideflash to a woman's arm from the kitchen sink, which is one of the metal bodies most likely to be charged with lightning current during a strike. Such sideflashes are usually not fatal. An exception is when current travels through the heart region, as is likely to be the case when a person serves as a bridge between two metal conducting paths.   Drawing by Valerie Frydenlund.

It ruptured electrical wiring at roof level, ignited a fire there, then travelled to shallow-driven galvanized ground rods on either side of the building. Although the victim was standing nearly 5 feet from one of the ground rods, lightning flashed off the down conductor and struck her at head level. She suffered severe trauma, amnesia, muscle weakness, and headaches, conditions that all are common to lightning victims after such incidents.

Traumatic and sometimes fatal flashovers to persons occur when lightning is denied an easy path or paths to adequately receptive earth. In the first instance cited, the tree's root system did not provide adequate grounding. The victims, whose feet were on a concrete slab with rainwater flooding across it, provided alternate conductive paths. In the second instance, grounding was inadequate due to sandy soil and an inadequate number of grounds which, in addition, were not driven deep enough into the earth.

Touch Voltage

Not infrequently, people are shocked and sometimes killed as they step in or out of automobiles or other vehicles when lightning discharges in the vicinity. Whether the jolt should be termed step voltage or touch voltage might be a matter of semantic judgement. However, it is generally agreed that *touch voltage* is the appropriate term when a voltage occurs in a generally vertical direction. A similar incident was once reported when an aircraft being refueled was struck, and the airline employee doing the refueling bridged the path to ground and was killed.

In other instances, the touch voltage designation is more clear cut. One example occurs when a person touches two different conductors or conducting surfaces carrying lightning current in a generally vertical direction. Another example is when someone touches the same current-carrying conductor at two different points, in a manner that allows the current to flow through the body or a part of it.

Step Voltage

The term *step voltage* refers to voltage gradients along the ground or other horizontal surface. The amount of step voltage a person might be subject to depends on the amount of current flowing along the ground, the depth to which the current through the ground penetrates or is conducted, the conductivity of the ground, and the distance between one foot and the other (see Fig. 5.9).

Let's examine an example where step voltage could be expected to produce three different results.

Lightning strikes an oak tree. One person is walking toward the tree and is 15 ft from it, taking a 3-foot stride. Another individual is standing about 5 feet from the trunk, with his feet at right angles to the current, which is flowing radially outward from the trunk. The second person is holding a horse by a rope, with the animal facing the tree so that its forefeet and hind feet are separated along a radial line from the tree trunk—the source of the voltage.

Oak trees have shallow, limited root systems, making them highly vulnerable to fatal injury by substantial lightning current. In this case we'll assume that the

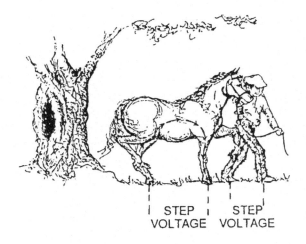

STEP VOLTAGE STEP VOLTAGE

FIGURE 5.9 Step voltage is electric current equivalent to the voltage drop between one leg and another. The man shown will likely escape with nothing more than temporary numbness and burn marks on his legs as the current travels from the foot nearest the tree, up that leg, and then down the other leg and back to ground. The horse is not as fortunate, for when ground current travels from the rear leg to the foreleg, it must pass through the heart region, almost certainly causing heart stoppage and, most likely, death. Drawing by Valerie Frydenlund.

result includes high current concentration at and near the ground's surface, a situation aggravated when the soil is laced with poorly conductive sand or gravel or is shallow due to the presence of bedrock near grade.

Let's assume that the lightning flash is a typical one, with current flowing at maximums of 25,000 A and 30,000,000 V of potential in four pulses, with a total flash duration of 0.4 seconds.

The person striding toward the tree would receive a very severe jolt through the lower torso. The step voltage sustained would be equal to the voltage drop between his feet, equated to the field strength times the 3-foot distance between them. The current flowing up the forward leg and down the rear leg would cause a powerful involuntary muscular reaction and, typically, the victim would relate, mistakenly, that he was "thrown through the air by lightning."

The individual standing 5 feet from the source of the voltage would emerge relatively unharmed from the incident, having been too distant to receive a sideflash. Because his feet were equally distant from the tree trunk, there would be very little, if any, voltage gradient between them. The lead rope (an insulator) would minimize current flow from the horse to the person.

The horse would take the brunt of the step voltage. He would suffer fatal effects of large current flows from forefeet to hind feet. There probably would be cardiac arrest as well as severe internal hemorrhaging from current flowing through its vital organs.

WHY LIGHTNING IS UNDERRATED

Perhaps the most exhaustive study of lightning casualties done in the United States was one by the National Oceanic and Atmospheric Administration (NOAA), covering a 34-year period ending on January 1, 1974. As reported in *NOAA* magazine, the study cited lightning as "the underrated killer."

The study showed that, during that period, lightning killed about 7,000 people in the United States as a result of direct strikes—many more than are killed by tornadoes, floods, or hurricanes. Author Edwin P. Weigel pointed out,

> This total . . . is undoubtedly conservative. . . . Many lightning victims are not included in national summaries because lightning usually kills only one person at a time. . . . Lightning deaths don't attract nationwide attention.
>
> Paradoxically, though, the statistics available indicate that the annual number of deaths in recent years have been much lower than death tolls earlier in this century, even though the U.S. population has increased greatly. There were, for example, 120 per year for the decade ending in 1973, compared to an annual average of 329 during the decade of the 1970s. This apparent drop is misleading.

Weigel then went on to explain that NOAA itself does not include lightning death tolls in its monthly periodical, *Storm Data,*

. . . because we are confident that any annual total of lightning deaths we would produce would be too low for the nation as a whole.

Statistics compiled by the Public Health Service prior to 1953 contain other causes of death, such as those resulting from fires caused by lightning. In 1953, Health Service coding rules were modified so that all deaths caused by an accident that was secondary to the effects of lightning were to be assigned to that cause rather than lightning.

The net result of all this is that deaths due to lightning strikes are often listed as due to such immediate causes as "electrocution," "smoke inhalation," and, in one particular case where an elderly woman died after lightning current ignited her television set, "carbon monoxide poisoning."

The author can attest to the accuracy of the Weigel report. For a number of years, when news clipping services were still available at moderate cost, the Lightning Protection Institute commissioned such services. The number of deaths due to lightning reported in those news clippings consistently topped the "official" reports by the National Weather Service, even though they only covered the months from April through September.

A study by the ESSA Atmospheric Physics and Chemistry Laboratory of lightning fatalities occurring over a seven-year period produced some interesting statistics:

- Seventy percent of all lightning deaths consist of single death due to a single discharge.
- Nearly 15 percent of all lightning deaths occur in groups of two, the balance being three or more deaths in a single event.
- About 70 percent of all injuries and fatalities due to lightning occur in the afternoon, 20 percent occur between 6:00 P.M. and midnight, 10 percent between 6:00 A.M. and noon, and about 1 percent between midnight and 6:00 A.M.
- While there were lightning-caused injuries in all of the lower 48 states during that period, Washington and Oregon reported no lightning fatalities, and Alaska and Hawaii reported neither lightning deaths or injuries.
- There is a higher than average incidence of lightning-caused fatalities along some of the nation's principal waterways—the Mississippi, Ohio, and Hudson Rivers, and their drainage basins.
- Other lightning-prone areas are along the Gulf Coast, particularly at Tampa, Florida, and in the mountains of Colorado.

Study Confirms Some Popular Notions . . .

Over a three-year period, the author conducted a detailed study of 1,000 lightning deaths and injuries covered by newspaper casualty reports. Resulting statistics confirm some existing beliefs about lightning as a peril, but call others into question. Fresh evidence seems to nudge some old wives' tales toward oblivion. Among the statistical affirmations are the following.

- The typical house is no lightning-safe haven; it is instead the site of 25.1 percent of all male lightning casualties and 61.0 percent of female deaths and injuries due to lightning.
- Men and boys are more foolhardy in exposing themselves to thunderstorm dangers than women and girls, and they pay the price with a three-to-one ratio of loss of life and limb.
- Women and girls, however, are more talkative in the teeth of peril, suffering three times as many phone-related lightning casualties as males.
- Fishermen who stand in open boats, and farmers who ride on tractors lacking metal cabs, are more often killed than injured when struck.
- Lone trees are deadly shelters during lightning storms.

. . . And Some Widely Held Notions Are Struck Down

The study revealed that three commonly held opinions are erroneous:

- *Misconception:* More golfers are killed by lightning than any other category of sports participants.
  Fact: Fishing is the deadlier sport, with twice as many deaths as golfing (although fewer injuries).
- *Misconception:* Lightning is largely a rural danger.
  Fact: Farming was once the most dangerous category, but farmers have dwindled drastically in number, and now farming ranks fourth and last behind (1) recreational; (2) industrial- commercial; and (3) residential.
- *Misconception:* You are completely safe from lightning inside a closed automobile.
  Fact: The study turned up deaths and injuries among people inside moving cars. In one, the rear window was smashed by lightning, and the driver, the only occupant, was injured. In another, lightning knocked a tree over a high-voltage power line, electrocuting the woman driver. A number of people have been temporarily blinded by lightning flashes, causing accidents and injuries.

Statistics From the Survey

Injuries "at home," which included outdoors and on the porch, nudged out "recreational" incidents by 39.6 percent to 28.0 percent of all *injuries*. But because the primary target was the house or a tree in the yard, there were fewer lightning *deaths* at home than in recreational pursuits. Table 5.1 lists the rankings of the categories of casualty sites.

TABLE 5.1   Lightning Deaths and Injuries by Categories of Activity (by percentage)

	All Deaths	All Injuries	Male Casualties	Female Casualties
Recreational	43.5	28.0	36.0	22.9
On farms	22.7	11.4	18.2	4.2
At home	20.4	39.6	25.1	61.0
Industrial/commercial	13.4	21.0	20.7	11.9

Among females, yards were most often death sites, and the yard was the most frequent site of injury. Among men, direct strikes in open locations were most frequent, and the most common site of injury was inside a building.

To get a clearer picture of lightning's worst danger spots, the author made a second breakdown, selecting the seven most common locations in which deaths or injuries occurred. All other types of locations were lumped in a "miscellaneous" category.

Table 5.2 shows, in descending order of frequency, lightning's danger spots.

TABLE 5.2   Lightning's Danger Spots

	All Deaths	All Injuries	Male Casualties	Female Casualties	All Casualties
Inside a building	19	217	142	94	236
In the open	92	135	202	25	227
Under a tree	41	50	77	14	91
In or on the water	51	39	84	6	90
In a tent or campsite	14	64	65	13	78
In or by a vehicle	37	40	71	6	77
Outside a building	10	41	44	7	51
Miscellaneous sites	49	101	79	150	15
Totals	313	687	764	236	1,000

QUESTIONS AND ANSWERS

Table 5.2 raises many questions of "why?" Some of these are expressed and answered below.

Q: Why are there three times as many male lightning casualties as female?

A: There are several reasons. Men and boys have a tendency to wait until the last moment before seeking shelter from the rain—and that is the most dangerous lightning period. Females are not as reluctant to show fear. More men do outdoor work. More boys play outdoors, although girls and women are catching up in the area of outdoor sports.

Q: OK, but why are there more male casualties than female, even indoors?

A: We dug deeper and found that, even indoors, men and boys tend to ignore safety rules, and perhaps also to be macho.

Inside the house, which was the only location with more female than male casualties (68 female to 47 male), there were 7 male casualties at open doors and windows, as opposed to only 3 female. The same greater male casualness was evident in the statistics for indoor casualties in both industrial/commercial and farm settings.

Many of the female at-home casualties were telephone related, showing that women and girls tend to spend more time on the telephone while at home than do men and boys.

Q: Why are there more deaths than injuries "in or on the water?"

A: When alone on a beach or in an open boat, a person is the highest point of concentration for positive ground-based ions. Thus, he is a target for a direct strike, which is almost always fatal unless competent aid is immediately available.

Q: If that is so, why aren't all "in the open" casualties fatalities, instead of 135 injuries to 92 deaths?

A: One reason is that many were multiple casualties among, for example, a group of workmen or team of boys on a football field. In most such cases, the person taking the direct strike is most often killed, while his companions are merely injured.

Another reason is that direct strikes are not always fatal, even when CPR or other aid is not available. One scientist pointed out that current tends to flow on the outside of the body in the case of a flash of high electrical magnitude. This "skin effect" is similar to a flow of electrical current along the outer surface of a conducting wire.

Dr. Cooper's Findings

An in-depth study of lightning-caused deaths and injuries, conducted by Dr. Mary Ann Cooper, was published as a chapter in the book, *Wilderness and Environmental Emergencies*. Dr. Cooper found that lightning victims can be placed in one of three groups, as follows.

1. Patients suffering minor consequences are awake and often report a feeling of being hit on the head or having been in an explosion. They are often confused and may suffer temporary deafness, blindness, or unconsciousness at the scene. They seldom suffer burns or paralysis but complain of a sensation of prickling or tingling of the skin, muscular pain, and confusion lasting from hours to days. Recovery is usually gradual and complete.
2. Victims suffering moderate injuries may be disoriented, combative, or comatose. They often suffer paralysis, with mottling of the skin and diminished or absent pulses, particularly in the lower extremities. Sometimes they suffer cardiopulmonary standstill. The heart, however, has a strong tendency to restart by itself, as opposed to the respiratory system. If CPR is not given promptly, the victim may suffer seizures and/or oxygen deficiency.

   First- and second-degree burns may appear during the first few hours. Membrane ruptures and basilar skull fractures sometimes occur as a result of the strike's powerful shock wave. Victims are prone to suffer sleep disorders, irritability, difficulty with fine psychomotor functions, and a general weakness.
3. Severely injured victims of lightning strikes may suffer cardiac arrest with either ventricle standstill or fibrillation. Cardiac resuscitation may not be successful unless very promptly administered. Direct brain damage may result from lightning's blast effect. Patients with other signs of blunt injury are likely to be the victims of a direct hit, although sometimes no burns will be seen. The prognosis is usually very poor in this group.

Dr. Cooper found that the most common cause of death in a lightning victim is cardiopulmonary arrest (stoppage of both the heart and breathing). In fact, a victim is highly unlikely to die unless cardiopulmonary arrest is suffered as an immediate effect of the strike. Nearly 75 percent of the people who suffered cardiopulmonary arrest from lightning injuries in the past died, many because CPR was not attempted.

The second major cause of death or severe and lasting injury is damage to the central nervous system. When lightning current traverses the brain, probable damage includes coagulation of the brain and intraventricular hemorrhage.

Dr. Cooper's studies indicate that 72 percent of lightning victims suffer loss of consciousness. Nearly 75 percent of those who do lose consciousness also suffer cardiopulmonary arrest. Victims sometimes suffer lasting sleep disturbances, storm neuroses, difficulty with fine mental and motor functions, and headaches.

Lightning Safety and the Law

Lightning safety is a personal matter when people are in houses or other buildings that they own, or in outdoor situations where nothing other than their own exercise of prudence is a practical response to lightning's dangers. But in public places, such as commercial or industrial buildings, government facilities, loading docks, construction sites, airports, golf courses, public parks, and other sites owned by others or by the public at large, lightning safety may be an important consideration. This is particularly likely to be true in areas where thunderstorms and lightning flashes to ground are frequent.

Lightning death or injury, traditionally viewed by the courts as an unavoidable "act of God," has been under assault by attorneys on behalf of victims or their survivors for several years. As a result, several judgments in favor of plaintiffs who claimed owner liability are on the books, placed there under the emerging premise that, "Yes, lightning is unavoidable, but because injury from it is avoidable, lightning protection is the responsibility of prudent men."

Personal Lightning Safety

In descending order of safety, recommended thunderstorm havens are:

1. buildings with proper lightning protection systems and nonconductive floors
2. large, steel-framed buildings
3. closed metal vehicles (but stay inside the vehicle, and either park or drive slowly)
4. large buildings without lightning protection systems (provided that you stay away from plumbing and other metal paths, open windows, doorways, and porches)
5. small, unprotected buildings such as houses (but, again, stay away from likely lightning paths and potential sites of flashover currents)

In descending order, most indoor lightning casualties occur while:

1. talking on the telephone, particularly in rural locations
2. in the kitchen
3. watching television
4. at a door or open window

Outdoors, avoid exposed shelters, open fields, open boats, lone trees, and large trees in groves. Alight from your golf cart, bike, or horse, and take cover. Don't swim. You're safer in a ravine, under a small tree in thick timber, in a ditch, or under a cliff. On the farm, avoid machinery and wire fences and dirt-floored, unprotected sheds and barns.

If you are caught in the open, kneel down and bend low, keeping only the feet and knees in contact with the ground. If someone near you is struck, give mouth-to-mouth artificial respiration, if you can detect the victim's pulse. If not, use heart compression as well. Have someone summon a rescue squad or emergency medical services immediately.

6

Lightning Fires and Destruction

Lightning current has the capability of inflicting five different damaging effects. Three of the effects usually are outcomes of direct strikes, but they may also be caused by airborne or conducted currents from quite distant flashes.

The three consequences that cause damage to buildings, trees (see Fig. 6.1), mechanical equipment, electrical equipment, and other objects in direct current paths are lightning's (1) thermal effect, (2) electrical effect, and (3) mechanical effect.

Indirect effects are (1) electric field induction and (2) magnetic field induction. These are of primary concern as causes of damage to electrical and electronic equipment. However, in the case of a nearby flash (particularly one of high electrical magnitude and duration), fire, mechanical damage, or casualty among persons or animals is possible as the result of induced currents.

LIGHTNING'S THERMAL EFFECTS

In layman's terms, there are "hot bolts" of lightning and "cold bolts." These terms are convenient designations, but they may be somewhat misleading, for two reasons.

FIGURE 6.1   Lightning's favorite target is a tall, lone standing tree or the tallest tree in a grove. Trees suffer damage according to their species, the time of year, and whether it is raining. For example, the heartwood of oak trees usually provides a moist and conductive path. Other trees are sometimes peeled or even completely shattered. "Hot bolts" of lightning ignite more forest fires than all of America's careless campers and smokers combined.   Photo by Cokato Enterprise, Cokato, MN.

First, typical cold-bolt strokes will generate temperatures near 50,000° F, but for so brief a duration that heat cannot build up appreciably in the conductors that are struck. A typical "hot bolt," on the other hand, is shown as having two component strokes, with the final stroke having a long tail of continuing current. The duration of heat in a cold bolt is measured in thousandths of a second, but it is measured in tenths of a second in a hot bolt.

The second reason is that lightning's targets vary greatly. Therefore, the result of a lightning flash is the sum total of the lightning's ability to inflict damage plus the particular target's vulnerability to the peculiar properties of that flash.

The big difference between "hot" and "cold" bolts, then, is not optimum temperature but duration, due to the hot bolt's atypical tail of continuing low amperage current. Obviously, a lightning flash composed of many leader/return stroke sequences will cause somewhat more heat buildup in struck objects' conducting paths than will brief-duration zaps composed of fewer pulses—but the difference is negligible. At least one stroke with a long, continuing tail of current is needed to generate enough heat to ignite wood. A 1967 study by D.M. Fuguay of seven flashes that caused forest fires showed that all seven contained a long-duration tail of current.

As indicated, structures vary greatly in vulnerability to lightning's thermal effect (see Figs. 6.1 through 6.3). The result of lightning current flowing through a conductor is determined by the current flow's amperage and duration and the material's electrical conductivity. If lightning strikes a poor conductor or insulator, the point of contact can be raised to a high temperature, which can result in fire, melting, burn-through, or shattering, depending on duration, the amount of heat transfer, and the type of material. The following sections provide examples of varying results.

Thermal Effect 1: Fire Due to Heat Transfer in Poor
Conductors

A study by the author of rural fires ignited by lightning during a three-year period indicated that certain classes of buildings of similar materials are more severely affected by lightning's thermal effect than others. Tall, old-fashioned barns were found to be very vulnerable. Among such barns said to have been ignited by lightning, 9 out of 10 burned to the ground. They were complete losses even when firefighters arrived promptly. Rural houses, normally of similar frame construction and also located distant from fire stations, were burned to the ground in about 3.5 cases out of 10.

Why the difference? Among the reasons are quicker discovery of fires in houses and easier access to residential fires with garden hoses and hand-held fire extinguishers.

FIGURE 6.2   The Grace Baptist Church of Brooklyn, New York, stood through nearly a century of thunderstorms until lightning set it afire. "I heard a terrific explosion," said a neighbor. "When I rushed to the window, I saw flames leaping out of the church and heavy smoke rolling down the block. The whole place was on fire." Six firemen were hurt battling the inferno caused as a "hot bolt" of current ignited a long path of fire that is typical of lightning flashes to flammable structures.   *New York Post* staff photo by Neil Schneider.

FIGURE 6.3 When tall, old fashioned barns are struck by "hot bolts" of lightning, they face a chance of 9 in 10 that they will burn to the ground. Usually highly flammable and too tall to be reached by garden hoses, such barns have been lost to rural America by the tens of thousands due to lightning strikes. Photo courtesy of Wally's Studio & Record Mart, Sheboygan Falls, Wisconsin.

FIGURE 6.4 Thunder crashed and lightning flashed to earth during an early spring snowstorm that blanketed the Elkhorn, Wisconsin, area. Among the targets of the storm's "hot bolts" were a metal garage/repair shop and a calf barn on the David MacLean farm. Photo by *Elkhorn Independent*.

But also prominent among the reasons is the fact that older, unprotected farm barns have few if any uninterrupted metal paths to ground. This results in instantaneous ignition of long paths of fire as lightning current travels through the barns' highly resistant, highly flammable roofing, siding, and framing materials.

Lightning fires among many unprotected houses, on the other hand, tend to start smaller and spread more slowly because they are more likely to be localized due to the presence of nearby alternate paths. Such paths include metal water lines, metal heating pipes and ducts, and metal plumbing stacks.

The Plastic Plumbing Factor

Modern, unprotected houses with plastic plumbing vent stacks are more likely to be ignited by lightning than those with metal plumbing systems. Metal plumbing stacks, grounded as they are to either municipal sewer systems or countryside underground waste disposal systems, serve in some degree to capture, conduct, and ground lightning current, particularly in instances of lightning strikes of minor electrical magnitude.

Buildings with plastic plumbing stacks, on the other hand, may be able only to present their electrical wiring systems as grounded metal paths. When that is the case, even a "cold bolt" of minor magnitude may cause a disastrous in-the-wall fire by shorting out wiring and causing an electrical fire.

Ironically, very modern houses with plastic plumbing lines often contain modern electrical and electronic systems and devices that are highly vulnerable to induced currents from nearby strikes as well as direct strikes. Such currents are usually directly induced into unprotected electrical wiring systems rather than entering through exterior electrical service lines where they could be controlled by surge suppression equipment.

Thermal Effect 2: Melting or Sparking Due to Heat Transfer in Metal

The potentially damaging effect of heat transfer at the point of contact of a lightning flash is often illustrated by pointed lightning rods whose tips are melted (sometimes slightly, and on occasion quite severely) when they receive lightning strikes. When lightning rods are properly grounded through multiple low-resistance conductors, the only negative effect of such melting is a possible reduction of the rods' heights to a point below the minimum height required by code. In a substandard system, on the other hand, high heat transfer may take place along the entire conductor, resulting in fire or other damage. The amount of heat transferred to metal is proportional to current amplitude, duration of the power dissipation, and the amount of electrical resistance due to the type of metal and its cross-sectional area.

Copper conductors are sometimes preferred in lightning protection as well as in electrical wiring, partly because of superior electrical conductivity. However, calculations indicate that the difference in conductivity is too small to be of concern. The temperature rise in an aluminum conductor is 1.5 times greater than in a copper conductor, and in steel the rise is 10 times greater than in copper of the same cross-sectional area.

As small-diameter metal wires are decreased in cross section, resistance to the passage of electric current increases proportionately (see Figs. 6.4 and 6.5). Excessive lightning currents will heat such wires by electron collisions with the metal lattice, and at some point the wires will melt or evaporate. Benjamin Franklin once observed that a lightning conductor ought to be at least the thickness of a goose quill for the sake of safe conductivity.

In lightning protection, however, adequate conductivity and cross-sectional dimension of conductors are not the only considerations. Bonds and joints also must provide good electrical contact and low resistance.

If bonds or joints initially, or as a result of corrosion or accumulation of poorly conductive debris, offer high resistance, excessive heat will develop, and ignition due to sparking may result. For that reason, overlapping metal roofing should never be used in lieu of lightning conductors, no matter how thick the sheets.

Thermal Effect 3: Melting of Metal Due to Burn-Through

Lightning protection standards require that metal cladding of less than 3/16 in. may not be substituted for lightning conductors, even though it is electrically continuous. That prohibition was written into code as a result of a combination of experience and calculations involving current amplitudes, durations, and resulting charge transfer.

A long-duration lightning flash can transfer hundreds of coulombs of charge, and the depth of penetration and/or size of the hole burned through is proportional to the amount of charge transfer. Incidents of burn-through in aircraft skins, metal roofing, and metal tank tops have underscored the critical importance of using lightning rods on thin-metal clad structures to serve as the points of current contact, particularly when there are explosive or flammable materials or vapors beneath the metal coverings.

Thermal Effect 4: Shattering of Poor Conductors

A different result of heat transfer occurs when heavy lightning currents travel through such poor conductors as trees, concrete, masonry, plastic, and lumber. Traces of moisture in such materials are evaporated and converted into high-pressure steam said to equal, in some instances, the explosive power of several kilograms of TNT.

FIGURE 6.5 Trouble down the line is the likely result when lightning strikes near or on a power line. The strike induces a large surge that quite often is not caught and grounded before it causes an overvoltage problem in a house or other building down the line.   Photo by Mac Overton, *Big Sandy* (Texas) *Journal.*

FIGURE 6.6 The surge described in Fig. 6.4 could cause an electrical line fault, but in this instance the most likely source of overvoltage was a lightning strike to the TV antenna beside the mobile home. Photo by Vincennes *Sun Commercial*.

Trees are lightning's most numerous and therefore most abused targets. The type of damage done to a struck tree depends on a number of factors. The tree's species dictates the extent and depth of its root system as a ground, as well as the amount and location of its moisture content. The conductivity of the soil at the site is a factor, as is the season of the year. If rain has wet the bark, current may flow on the outside of the trunk, minimizing damage. If not, lightning is likely to follow the moist heartwood or the moist area beneath the bark.

Forest fires are most numerous during dry seasons, and the tops of specimen trees are most often ignited by lightning during dry fall periods. Estimates of the percentage of trees that die as a result of lightning strikes vary. Many are struck by low-magnitude flashes that leave little if any evidence of the strike. In severe damage instances, lightning current travels through the moist areas, and high pressure steam produces various types of damage. Oak trees are often split, the bark of pine trees is often peeled, and elms, with their tightly woven grain, are sometimes splintered into pieces.

After trees, the next most numerous (and therefore most common) victims of lightning's shattering effect are unprotected masonry chimneys and smokestacks. Large charge transfers from lightning flashes heat and vaporize traces of moisture in the chimney's mortar joints. This, with harmonic vibration, causes an explosive outward thrust that may hurl bricks across wide areas. Concrete block walls are likely to sustain cracks along mortar joints when struck by lightning. On occasion, they are toppled as a result of the damage.

A common example of lightning damage is spalling or cracking of concrete, particularly at parapets and the top edges of such structures as unprotected concrete silos. Fiberglass church steeples, being very poor conductors, have been shattered, split, or otherwise damaged by lightning so often that manufacturers of such steeples have begun to equip them with lightning rods and concealed downconductors.

Wood-framed structures often sustain "cold-bolt" damage as a result of the shattering effect of lightning current on poorly conductive materials containing traces of moisture. The effect is enhanced by the additional force exerted by harmonic vibration.

LIGHTNING'S ELECTRICAL EFFECT

When a heavy flow of rapidly varying current is introduced into a lightning conductor, the inductance of the conductor may combine with a strong resistance at ground level and sufficient current flow duration to produce a spark to another nearby grounded object. This "sideflash" is the result of lightning's need to establish additional paths on which to unload high induced voltages that exist for about one microsecond at the beginning of each return stroke.

Flashover frequency and distance are dependent on building height, which determines the length of the conductor and, therefore, the amount of voltage induced into it, the number of conductors present to carry current downward, and the amount of resistance encountered at ground level. Since those factors influence flashover, prevention of such sparking calls for measures to decrease induction, increase conductivity, and lower ground resistance.

LIGHTNING'S MECHANICAL EFFECT

It is lightning's mechanical effect that causes thunder and which, on occasion, causes walls to topple, roofs to cave in, and small structures to be literally blown apart. Thunder is the result of a sound wave. A sound wave, in turn, is produced at distances of a few meters around a cylindrical shock wave, which is produced by rapid heating of air that surrounds the lightning channel.

A pressure of several atmospheres is produced at the lightning channel, and we hear the resulting sonic boom as thunder. The shock wave diminishes quickly as it travels outward from the channel. Therefore, it is important to keep some distance between the lightning strike point and other surfaces or objects that can be damaged by violent force.

R. D. Hill reported the results of his studies on "Channel Heating in Return-Stroke Lightning," published in the *Journal of Geophysical Research* (January 1971). Using as a basis for calculation a 30,000 ampere return stroke, Dr. Hill showed a pressure of 40 atmospheres at 0.75 cm distance; 29 atmospheres at 2 cm, and 7 atmospheres at 4.1 cm.

While Hill's conclusions are theoretical, they are valuable additions to lightning protection technology and assist in explaining lightning's mechanical effects. The following are two examples of those effects.

1. A man is walking toward a tree as it is struck by a lightning flash. He later recounts that he was "thrown through the air." In truth, he was lifted upward by a combination of lightning's "blast effect" and the involuntary reaction of powerful leg muscles to "step voltage," as explained in the preceding chapter.
2. A chimney is struck and its upper portion demolished (see Figs. 6.6 through 6.8). Neighbors find bricks from the chimney in their yards. Here again, two forces are at work. First, tremendous heat creates expanding steam from traces of moisture in mortar joints. Second, the blast effect of current that tends to flow through the carbon sheet on the inside surface of the chimney gives an extra boost to the loosened bricks.

Another source of mechanical damage is magnetic interaction of lightning current. The result is a physical force applied to a conductor through which current is

FIGURE 6.7   The carbon path inside a chimney and traces of moisture in mortar joints both provide lightning current paths that are highly resistant yet better than nonconductive air. Bricks are sometimes hurled throughout a neighborhood by lightning's blast effect.   Photo by Joe Romeburg, *Democrat Messenger*, Waynesburg, PA.

FIGURE 6.8   Fortunately, school was out when a summertime lightning strike dislodged concrete and masonry that crashed through the roof into this schoolroom.   Photo by the *Goshen News*, Goshen, IN.

traveling. This force will tend to straighten out a conductor in the same way that high-pressure water tends to straighten a hose. It is because of the mechanical effect produced by lightning's magnetic field that lightning protection code prohibits sharp bends in conductors and reentry loops, requires that conductors be strongly fastened at intervals of 3 feet or less, and requires certain minimum pull-out resistances for masonry and concrete anchors.

The existence of these magnetic forces also dictates that lightning conductors in parallel not be placed close to each other. The force that will be produced between parallel straight conductors is proportional to the square of the current and inversely proportional to the distance between the conductors.

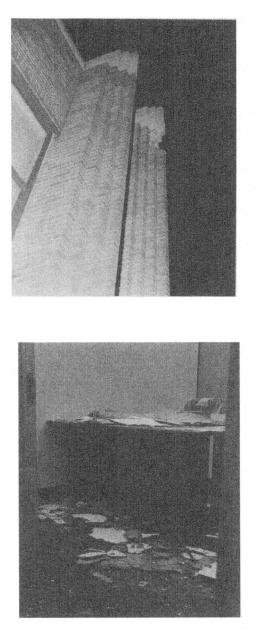

a. Strike Point

b. Dislodged Masonry

FIGURE 6.9   Lightning does not always strike near the tops of buildings. In this instance, a bolt struck near the top of the structure (a), dislodging masonry that crashed through the roof of a small, one-story attached structure (b). Fortunately, the structure was unoccupied at the time. Photo by Call-Chronicle Newspapers, Allentown, PA.

INDUCED EFFECTS OF LIGHTNING FLASHES

Chapter 4 describes the changes that take place in atmospheric and earthbound electric fields during thunderstorm formation. During a lightning discharge, instantaneous changes in the electric field occur which may be as great as 15 kW at a distance of 3 miles.

In the case of a nearby lightning flash, structural damage can be caused as the result of powerful currents in the ground due to the presence of large electric and magnetic fields.

Electric Field Induction

Electric charge changes during lightning discharges induce current flows on conductors. These charges are proportional to the strength of the electric field and appear as voltages across resistances in conductors. As the electric field changes rapidly, current flows across the conductor impedances vary accordingly to maintain a balance with the field.

Magnetic Field Induction

When lightning strikes in the vicinity of a conductor that forms, or is part of a closed loop, a current will be induced to flow in the circuit due to the presence of a time-varying magnetic field. The voltage induced in the loop by this effect is proportional to the rate of change of the magnetic flux density and inversely proportional to the resistance within the circuit.

7

Lightning Risk Assessment

During the 1977 meeting of the National Fire Protection Association's lightning protection code committee, chairman John McGinnis (of Kemper Insurance) appointed a subcommittee headed by Russell Mattson (of Factory Mutual Research Association) to write a section on lightning risk evaluation. This author, then secretary of the code committee, sat on Mattson's subcommittee.

After reviewing British, Australian, and South American risk evaluation guides as printed in these countries' respective standards, the subcommittee presented its text, and the guide was printed as Appendix I of the NFPA code, 1980 edition.

The following material (with the exception of the examples of risk evaluation) is reprinted with permission from *NFPA 78-1989, Lighting Protection Code,* copyright ©1989, National Fire Protection Association, Quincy, PA 02269. (Table and figure numbers correspond to those of the original document.) This reprinted material is not the complete and official position of the National Fire Protection Association on the referenced subject, which is represented only by the standard in its entirety.

THE GUIDE'S ROLE

The risk assessment guide is not intended to be the sole source in determining the risk of loss to lightning. For one thing, it is not possible to cover each special design factor that may render a structure more or less susceptible to lightning damage. In special cases, personal or economic factors may be very important and should be considered in addition to assessment obtained by use of this guide.

If the structure in question is in a high-risk situation, a risk index (R) should be computed for a wide range of structures in the environment concerned. The structure then may be compared to the indexes of the other structures so that a judgment of local risk weighting can be made.

Determining the Risk

The assessment of risk index (R) is given in Table 1-2. The index is obtained by dividing the sum of the values given in Tables 1-2a through 1-2e by the lightning frequency index value obtained from Table 1-2f.

The risk index (R) is:

$$R = \frac{A + B + C + D + E}{F} \tag{7.1}$$

TABLE 1.2a   Index "A"—Type of Structure

Structure	Value
Single family residence less than 5,000 sq ft (456 m^2)	1
Single family residence over 5,000 sq ft (456 m^2)	2
Residential, office, or factory building less than 50 ft (15 m) in height	
Covering less than 25,000 sq ft (2,323 m^2) of ground area	3
Covering more than 25,000 sq ft (2,323 m^2) of ground area	5
Residential, office or factory building 50 to 75 ft (15 to 23 m) high	4
Residential, office or factory building 75 to 150 ft (23 to 46 m) high	5
Residential, office or factory building 150 ft (46 m) or higher	8
Municipal services: fire, police, water, sewer, etc.	7
Hangars	7
Power generating stations, control telephone exchanges	8
Water towers and cooling towers	8
Libraries, museums, historical structures	8
Farm buildings	9
Golf shelters and other recreational centers	9
Places of public assembly such as schools, churches, theaters, stadiums	9
Slender structures such as smokestacks, church steeples and spires, control towers, lighthouses, etc.	10
Hospitals, nursing homes, housing for the elderly or handicapped	10
Buildings housing manufacturing, handling, or storage of hazardous materials	10

TABLE 1.2b   Index "B"—Type of Construction

Structural Framework	Roof Type*	Value
Nonmetallic (other than wood)	Wood	5
	Composition	3
	Metal—not continuous	4
	Metal—electrically continuous	1
Wood	Wood	5
	Composition	3
	Metal—not continuous	4
	Metal—electrically continuous	2
Reinforced concrete	Wood	5
	Composition	3
	Metal—not continuous	4
	Metal—electrically continuous	1
Structural steel	Wood	4
	Composition	3
	Metal—not continuous	3
	Metal—electrically continuous	1

*Note: Composition roofs include asphalt, tar, tile, slate, etc.

TABLE 1.2c   Index "C"—Relative Location

Location	Value
Structures in areas of higher structures:	
Small structures—covering ground area under 10,000 sq ft (929 m^2)	1
Large structures—covering ground area over 10,000 sq ft (929 m^2)	2
Structures in areas of lower structures:	
Small structures—covering ground area under 10,000 sq ft (929 m^2)	4
Large structures—covering ground area over 10,000 sq ft (929 m^2)	5
Structures up to 50 ft (15.2 m) above adjacent structures or terrain	7
Structures more than 50 ft (15.2 m) above adjacent structures or terrain	10

TABLE 1.2d   Index "D"—Topography

Location	Index Value
On flat land	1
On hill side	2
On hill top	4
On mountain top	5

TABLE 1.2e   Index "E"—Occupancy and Contents

	Index Value
Noncombustible materials—occupied	1
Residential furnishings	2
Ordinary furnishings or equipment	2
Cattle and livestock	3
Small assembly of people	4
Combustible materials	5
Large assembly of people—50 or more	6
High value materials or equipment	7
Essential services—police, fire, etc.	8
Immobile or bedfast persons	8
Flammable liquids or gases—gasoline, hydrogen, etc.	8
Critical operating equipment	9
Historic contents	10
Explosives and explosive ingredients	10

TABLE 1.2f   Index "F"—Lightning Frequency Isokeraunic Level

(from Isokeraunic Map)	Index Value
0–5	9
6–11	8
11–20	7
21–30	6
31–40	5
41–50	4
51–60	3
61–70	2
Over 70	1

Table 1.2f is derived from an isokeraunic map of the continental United States. Such maps are readily available for other parts of the world. The annual number of thunderstorm days is recorded as the total local calendar days during which thunder is heard, and since 1894 it has been defined in this manner. A day with any thunderstorms is recorded and counted as one, regardless of the number of thunderstorms occurring on that day. The occurrence of lightning without thunder is not recorded as a thunderstorm. (Data supplied by Environmental Science Service Administration, U.S. Department of Commerce.)

The following examples illustrate how the NFPA guide works.

Example 1—A Wood-Framed, Brick-Sided, Composition-Roofed Building Supply Center Located in the Chicago Suburb of Palos Heights

INDEX

A	Type of Structure	
	Commercial building over 25,000 sq ft, under 50 ft high	5
B	Type of Construction	
	Wood frame, composition roof	3
C	Relative Location	
	Large structure, covering ground area over 25,000 sq ft	5
D	Topography	
	On a hillside	2
E	Occupancy and Contents	
	Combustible materials	5
F	Isokeraunic Level	
	41 to 50 thunderstorms per year	4

Using these values with Eq. (7.1), yields:

$$R = \frac{5 + 3 + 5 + 2 + 5}{4} = 5$$

Referring to Table 2-2 of NFPA Assessment of Risk, R values between 4 and 7 have an overall risk rating expressed as moderate to severe.

Example 2—A Steel-Framed, One-Story Factory near Orlando, Florida

INDEX

A	Commercial building less than 50 ft high	5
B	Steel frame, composition roof	3
C	Large structure, over 25,000 sq ft	4
D	On flat land	1
E	Containing combustible materials	5
F	Over 70 thunderstorm days per year	1

The result using the Eq. (7.1) is:

$$R = \frac{5 + 3 + 4 + 1 + 5}{1} = 18$$

However, Florida is in the southeast section of the country, which happens to have a lower than average ratio of cloud-to-ground lightning flashes, as opposed to cloud-to-cloud flashes. Therefore, the R value 18 is multiplied by a reduction factor of 0.5, producing a final R value of 9, expressed as a severe risk.

Example 3—A Small Hardware Store Office/Display Building in Baltimore, Maryland

INDEX

A	Commercial building less than 50 ft high and covering less than 25,000 sq ft	3
B	Structural steel frame, composition roof	3
C	Low structure among higher structures, and less than 10,000 sq ft	1
D	On flat land	1
E	Containing combustible materials	5
F	11 to 20 thunderstorm days per year	7

The result using Eq. (7.1) is:

$$R = \frac{(3 + 3 + 1 + 1 + 5)}{7} = 2$$

However, northeastern seaboard states have a higher than average incidence of cloud-to-ground lightning flashes. Therefore, the R value is multiplied by a factor of 1.5, producing a final R value of 3, expressed as a light to moderate risk.

The range of risk values in the guide is "light," "light to moderate," "moderate," "moderate to severe," and "severe." The examples given illustrate that the presence of combustible materials such as paint, turpentine, roofing tar, and gasoline or kerosene push structures built of fire-resistant materials into medium or high lightning risk ranges. When a multiple-structured commercial, industrial, or institutional building is rated, the lightning risk value is usually considered higher than any single building in the complex due to the greater area involved.

8

Terms, Definitions, and Materials

As in other trades, professions, and business areas, lightning protection has its own glossary of terms and definitions as well as special tools and materials. A full-line manufacturer of lightning protection system components for structures may stock as many as 1,000 items. There are dozens of air terminal configurations, for example. And when surge suppression equipment is added, lightning protection equipment inventories expand greatly.

TERMS AND DEFINITIONS

Many of the terms and definitions that follow conform to the wording in the *NFPA 78* code. Where that is the case, they are included by permission from the National Fire Protection Association. Two terms deserve particular attention because they are inclusive of many others:

1. *Lightning Protection System.* A lightning protection system is a complete system of air terminals, conductors, ground terminals, interconnecting (bonding) conductors, surge suppressors, and other connectors and fittings to complete the system.

2. *Rods and Points.* These are generic terms for component parts. The correct term for the uppermost components is *air terminals,* which are composed of *rods* and *bases* as described in the following definitions.

Lightning Protection Terms

Air Terminal. A solid or tubular perpendicular rod, of specified size and material, provided with a mounting base having a conductor connection, intended to capture a lightning flash.

Bonding System. A system that is intended to establish electrical continuity between two bodies.

Cable. A conductor formed by a number of wires stranded together.

Chimney. A small concrete, masonry, or metal vent stack with a flue area of less than 500 in^2 (0.3 m^2) and a height of less than 75 ft (23 m), protruding through the roof or attached to the side of a building. A chimney typically is used in heating or venting and is subject to protection under Class I requirements.

Class I Materials. All conductors, fittings, and fixtures necessary to protect ordinary buildings and structures not exceeding 75 ft (23 m) in height.

Class II Materials. All conductors, fittings, and fixtures necessary to protect ordinary buildings and structures exceeding 75 ft (23 m) in height.

Combustible Liquid. A liquid that has a flash point at or above 100° F (37.8° C).

Conductors. (1) Those portions of a lightning protection system designed to carry current between air terminals and ground. (2) Bonding conductors used to accomplish potential equalization between ground metal bodies and lightning protection components.

Copper-Clad Steel. Steel with a coating of copper bonded to it, typically used for grounding devices.

Counterpoise (ground). A conductor encircling a building and interconnecting all ground terminals.

Down Conductor. A conductor that connects an air terminal with an earth terminal.

Fastener. A device used to secure to the conductor to the structure.

Flame Protection. Self-closing gage hatches, vapor seals, pressure-vacuum breather valves, flame arresters, or other reasonably effective means to minimize the possibility of flame entering the vapor space of a tank.

Flammable Air-Vapor Mixtures. When flammable vapors are mixed with air in certain proportions, the mixture will burn rapidly when ignited. The combustion range for ordinary petroleum products, such as gasoline, is about 1.5 to 7.5 percent by volume, with the remainder being air.

Flammable Liquids. Liquids that emit vapors that are flammable when mixed with air.

Flash Point. The flash point of a liquid shall mean the minimum temperature at which it gives off vapor in sufficient concentration to form an ignitable mixture with air near the surface of the liquid within the vessel as specified by appropriate test procedure and apparatus.

Gastight. Structures so constructed that gas or air can neither enter nor leave the structure, except through vents or piping provided for the purpose.

Ground Terminal. The portion of a lightning protection system extending into the earth (such as a ground rod, ground plate, or the conductor itself) serving to bring the lightning protection system into direct contact with the earth.

Grounded. Electrically attached to earth either directly or through a conductor.

High Rise Building. For purpose of lightning protection, a building over 75 ft (23 m) in height.

Metal-Clad Building. A building with sides and/or roof made of or covered with sheet metal.

Primary Metal Body. A metal object on or above roof level, not in a zone of protection, that is subject to a direct strike.

Secondary Metal Body. A metal body within a zone of protection that is subject to a sideflash due to a potential buildup that is opposite to that of a nearby conductor.

Shall. A mandatory specification.

Should. A recommendation.

Sideflash. An electrical spark occurring between metallic objects or from such objects to the lightning protection system or to ground.

Spark Gap. An air space between two conductors.

Stack, Heavy-Duty. A smoke or vent stack is classified as "heavy duty" if the cross-sectional area of the flue is greater than 500 in^2 (0.3 m^2) and the height is greater than 75 ft (23 m).

Striking Distance. The distance over which a final breakdown of the stroke to ground or to a grounded object occurs.

Vapor Openings. These are openings through a tank shell or roof above the surface of the stored liquid. Such openings may be provided for tank breathing, tank gaging, fire fighting, or other operating purposes.

Zone of Protection. The zone of protection provided by a grounded air terminal, mast, or overhead ground wire is that adjacent space which is substantially immune to direct strokes of lightning

Glossary of Materials

This guideline covers the material and manufacturing criteria for equipment and components used in lightning protection systems for conventional-type buildings. Components and equipment are generally grouped into two classifications: Class I (for ordinary structures 75 feet or less in height) and Class II (for ordinary structures more than 75 feet high).

All units of measurement are given in standard American units and gauges. Metric equivalents can be computed as required, but in no case shall decimal rounding or moves to metric industry standards result in a value less than the one cited herein.

The following pictorial index/glossary gives illustrative examples of the material types and components covered by this guideline and the terminology in their designation and description. The illustrations shown are for graphic description only and are not to be taken as requirements in and of themselves.

Chapters 9, 10, 11, and 14 address in detail the various components and applications of lightning protection systems.

Class I Components

Conductors for Class I lightning protection systems are copper or aluminum of a grade normally used for electrical conductors having 99+ percent purity and soft drawn in temper. Conductors may be of construction known as braided, rope-lay, or braided/twisted type (see Fig. 8.1). Main size conductors for copper and aluminum cables are shown in Table 8.1.

Additional types of acceptable conductors for Class I main conductors are solid rods or solid strips of copper or aluminum. Requirements for these types are also shown in Table 8.1.

Secondary size (miniature) conductors for Class I equipment are of copper or aluminum. Construction is of braided/twisted type, solid wire, or solid strip. Size and weight requirements are as given in Table 8.2.

Air terminal points (see Figs. 8.2 and 8.3) for Class I systems are of three basic types: (a) plain threaded-end type, (b) adaptor-end type, and (c) integral threaded hub type and shall be made of aluminum or copper rod, aluminum or copper tube, or cast aluminum or cast bronze of a size/thickness as outlined in the Lightning Protection Institute Standard LPI 175.

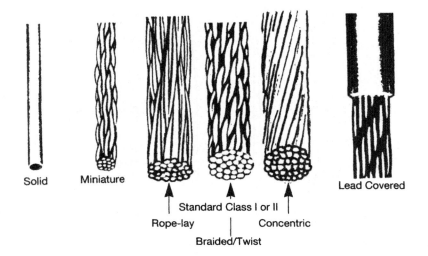

Solid   Miniature

Standard Class I or II

Rope-lay   Concentric

Braided/Twist

Lead Covered

FIGURE 8.1   Typical Lightning Conductor Types.

TABLE 8.1   Minimum Material Requirements for Class I Main Conductors

Type of Conductor		Copper		Aluminum	
		Standard	Metric	Standard	Metric
Cable	Min. size ea. strand	17 AWG	1.15 mm	14 AWG	1.63 mm
	Wt. per 1,000 ft	187.5 lb	85 kg	95 lb	43 kg
	Cross-sect. area	59,500 cm	.30 cm^2	98,500 cm	.499 cm^2
Solid strip	Thickness	14 AWG	1.63 mm	12 AWG	2.05 mm
	Width	1 in	25.4 mm	1 in	25.4 mm
Solid bar	Wt. per 1,000 ft	187.5 lb	85 kg	95 lb	43 kg
Tubular bar	Wt. per 1,000 ft	187.5 lb	85 kg	95 lb	43 kg
	Min. wall thickness	.032 in	.815 mm	0.0641 in	1.63 mm

TABLE 8.2   Minimum Material Requirements for Class I Secondary Conductors

Type of Conductor		Copper		Aluminum	
		Standard	Metric	Standard	Metric
Cable	Wire size	17 AWG	1.15 mm	14 AWG	1.63 mm
	Number of wires	14	14	10	10
Solid strip	Thickness	16 AWG	1.29 mm	16 AWG	2.05 mm
	Width	0.5 in	12.7 mm	0.5 in	12.7 mm
Solid rod	Wire size	6 AWG	4.12 mm	4 AWG	5.19 mm

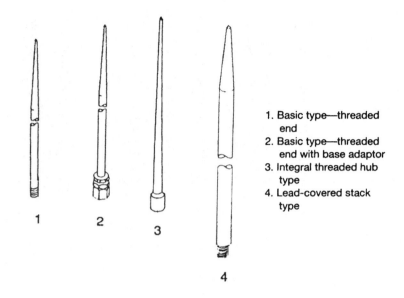

1. Basic type—threaded end
2. Basic type—threaded end with base adaptor
3. Integral threaded hub type
4. Lead-covered stack type

FIGURE 8.2 Typical Air Terminal Points.

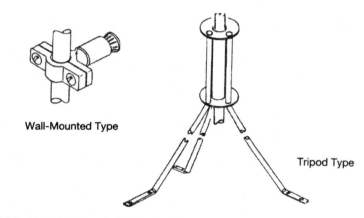

Wall-Mounted Type

Tripod Type

FIGURE 8.3 Typical Air Terminal Support and Brace.

Air terminal bases (see Fig. 8.4) must be adequate to support the point and provide a strong continuous connection to the connecting cable (see Fig. 8.5). Bases may be of copper, aluminum, or stainless steel of case or stamped construction.

Air terminals may be attached to the base by male threads to a female hub on the base or by means of a female hub or adapter on the point to a male stud on the

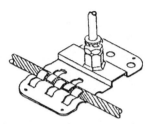

Crimp Type: Class I Only

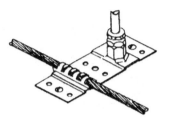

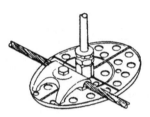

Cast Type: Class I or II

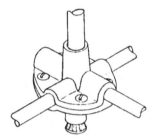

Lead Covered: Class II

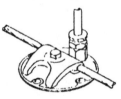

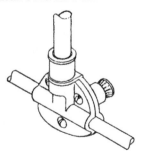

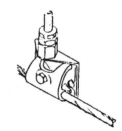

FIGURE 8.4   Typical Air Terminal Bases.

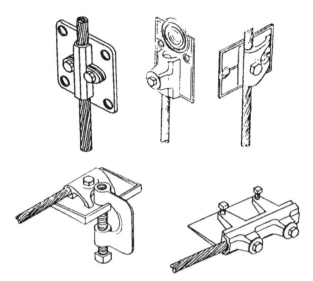

FIGURE 8.5   Typical Connectors for Bonding to Steel Used as System Element.

base. In either case, five full threads should be engaged on both the point and the base. The standards LPI 175 (Lightning Protection Institute) and UL96 (Underwriter's Laboratories) specify detailed requirements for all lightning protection components from air terminals, grounding devices (see Fig. 8.6), and conductors, to nails, clips, and screws.

Class II Components

Conductors for Class II main and secondary sizes should be of the same materials and construction as for Class I except that (a) sizes are increased to the minimums shown in Table 8.3, and (b) concentric-type construction is permitted for Class II aluminum main size cables only.

Air terminal points used in Class II installations are similar in the types of materials and construction to those used in Class I systems, except that tubular points are not acceptable, and sizes are increased from three-eighths inch (0.375") diameter to one-half inch (0.5") for copper air terminals, and five-eights inch (0.625")

TABLE 8.3   Minimum Material Requirfements for Class II Conductors

Characteristic	Copper		Aluminum	
	Standard	Metric	Standard	Metric
Minimum wire size	15 AWG	1.45 mm	13 AWG	1.83 mm
Weight per foot	6 oz	170 gr	3 oz	86 gr
Weight per 1,000 ft	375 lb	170 kb	190 lb	86 kg

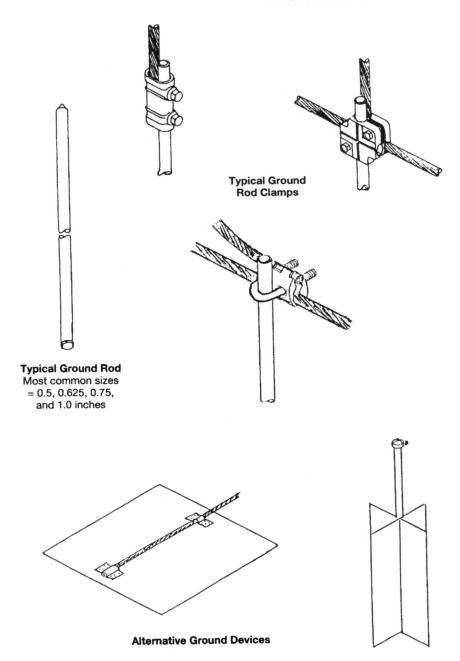

**Typical Ground
Rod Clamps**

Typical Ground Rod
Most common sizes
= 0.5, 0.625, 0.75,
and 1.0 inches

Alternative Ground Devices

FIGURE 8.6 Typical Ground Devices.

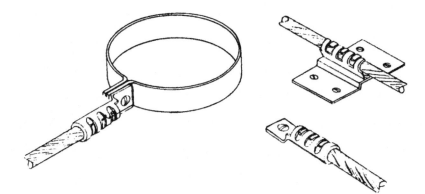

FIGURE 8.8 Bonding Devices, Crimp Type: Main Size—Class I Only.

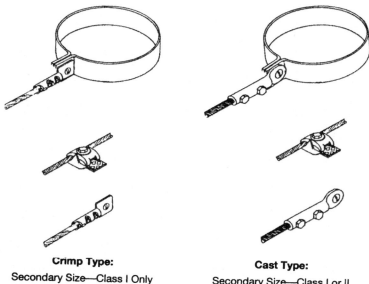

Crimp Type:
Secondary Size—Class I Only

Cast Type:
Secondary Size—Class I or II

FIGURE 8.9 Bonding Devices, Crimp and Cast Types.

diameter for aluminum air terminals; air terminal bases are of the same construction material and size as for Class I, except that no crimp-type cable connections are permitted (see Fig. 8.7). Bonding devices for Class II systems must have bolt pressure fittings; no crimp-type devices are permitted as in Class I (see Figs. 8.8 through 8.10).

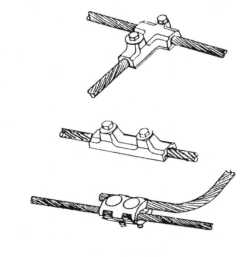

Crimp Type:
Main Cable—Class I Only

Bolt Pressure, Cast Type:
Main Size Cable—Class I or II

Crimp Type:
Secondary Cable—Class I Only

Bolt Pressure, Cast Type:
Secondary Cable—Class I or II

FIGURE 8.7   Typical Cable Splicers.

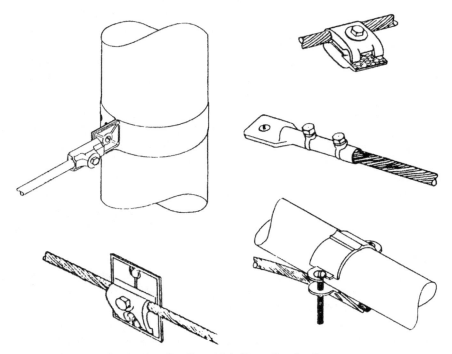

FIGURE 8.10   Bonding Devices, Cast Type: Main Size—Class I or II.

Grounding electrodes for Class II protection must be five-eighths inch in diameter, as opposed to the one-half inch minimum diameter permitted for Class I installations.

Class II Smokestack Materials

Conductors for heavy-duty smokestacks are copper only and of the same size as for Class II. All conductors on the upper 25 feet of the stack must have a minimum of one-sixteenth inch (0.0625") lead covering, which results in a cable weighing 900 lb per 1,000 feet. No secondary size cables are permitted anywhere on a smokestack.

Air terminals for such stacks must be of copper or stainless steel with a minimum diameter of five-eighths inch. Copper points must be lead covered. Stainless steel points should be alloy 302-304 and need not be lead covered.

Grounding devices for heavy duty stacks are the same as for all Class II structures. Bonding plates for connection to steel hoods, liners, and breechings are the same as for Class II bonding devices to steel frames and must be of copper, bronze, or stainless steel. Exposed copper or bronze items used on the upper 25 feet of the stack must be lead covered.

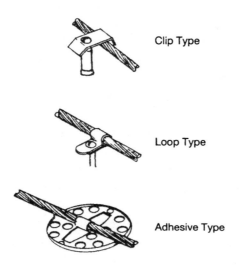

Clip Type

Loop Type

Adhesive Type

FIGURE 8.11   Typical Cable Supports.

Miscellaneous Components

Bimetallic fittings for joining copper and aluminum conductors or components are so constructed, using stainless steel isolators and/or electrically bonded bimetal shims that exclude moisture, so as to prevent electrolytic interaction of the two metals. Crimp-type fittings are permitted on Class I applications, but only bolt pressure devices are used for Class II applications.

Conductor guards for either Class I or Class II may be of pressure treated wood, PVC conduit, copper tube, galvanized iron, or formed galvanized steel, and are 8.0 feet long. Each guard must have at least two straps or six bolted or screw supports (see Fig. 8.11), depending on the type of guard used.

9

Capturing the Lightning Flash

A lightning protection system has a single, overriding purpose—to shield a building, its occupants, and its contents from the thermal, mechanical, and electrical effects described in accompanying pages. To do this effectively, the system must (1) capture the lightning current, (2) lead the current downward and, preferably, along the outside of the target, (3) prevent the occurrence of sideflashes both inside and outside the structure, and (4) conduct the current to a grounding system that (5) is able to accept the heavy flow of current and cause its harmless dissipation into the earth that surrounds the structure.

In times past, lightning protection component manufacturers and system installers promoted a second purpose, claiming that the occurrence of a lightning strike in the vicinity of the structure might be prevented by "bleeding off" ground charges into the atmosphere. Advertisements and promotional literature illustrated this alleged function along with capture, conduction, and grounding.

But atmospheric scientists' experiments, as well as many strikes to protected structures, have shown that while ground currents do indeed "bleed off" the tips of lightning rods, these currents are eventually carried off by gusts of wind and immediately replaced by new "point discharge" currents.

For the above reasons, strike prevention has been dismissed by atmospheric scientists as a practical function of code-adherent lightning protection systems. It has been pointed out that if it were possible for lightning rods to prevent lightning from striking, then it would be possible for trees and a myriad of other vertical projections to keep lightning from striking.

It is important, therefore, to realize that the only viable objective of a code-compliant system is to capture and then convey lightning current harmlessly to earth. When that objective is kept in mind, the principles that govern the design, installation, and inspection of lightning protection systems are clarified.

STRIKES TO UNPROTECTED STRUCTURES

It is helpful to visualize the functions of a lightning protection system's capture subsystem and to review the lightning-strike events described in Chapter 4. As a leader stroke descends, its stream of free electrons pauses at the end of each step, then continues, perhaps in an altered direction, neutralizing pockets of positive ions along the way. Several incomplete ionized channels may branch out from the main electron flow.

When the stepped leader's main channel reaches striking distance above effective earth (which may be the top of a building, a pole, a person, or another object), the discharge path becomes quite predictable. Upward-straining point-discharge streamers, which began to appear as the electric field intensified, propagate toward the downcoming stepped leader (see Fig. 9.1). Eventually, a successful positive streamer shoots up to meet the negative leader.

Where Lightning Strikes Houses

Which point on an unprotected building will launch the successful positive streamer and thus determine where the building will be struck? The author made a three-year study of newspaper clippings, selecting only those reports where the exact point of the lightning strike was known. The results of the study are shown in Fig. 9.2.

Private residences are by far the most numerous type of structures in the United States, and the profile of the strike points illustrated in Figure 9.2 probably provides the most comprehensive report available for lightning strikes to small structures. No similar statistics are known to be available for larger structures. However, the Empire State Building, which is 1,265 feet tall, was the subject of a lightning strike count for a period of about 10 years. Instruments stationed some distance away from the structure recorded an average of 23 strikes annually to the building.

FIGURE 9.1 Positive ground-based streamers strain upward off sharp high points as a negative "stepped leader" stroke nears ground.

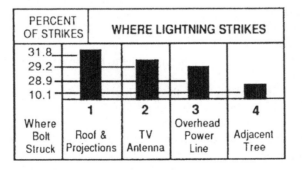

PERCENT OF STRIKES	WHERE LIGHTNING STRIKES			
31.8				
29.2				
28.9				
10.1				
Where Bolt Struck	**1** Roof & Projections	**2** TV Antenna	**3** Overhead Power Line	**4** Adjacent Tree

FIGURE 9.2 According to the author's study of lightning strikes to residences where the owner, a fire chief, or other individual fixed the strike point, corners and projections on the roof were most often struck, followed by rooftop antennas and overhead power lines.

Triggered Lightning Strikes

An interesting outgrowth of the Empire State Building study was confirmation of several facts:

1. Tall buildings "trigger" lightning strikes.
2. The strike radius of lightning increases in direct proportion to the magnitude of the flash.
3. Tall structures without lower roofs, balconies, or other projections are sometimes struck on the side by low-intensity flashes, with no resulting damage.
4. Tall buildings provide a certain amount of shielding for lower structures, but the shielded area is limited.

A DISCREDITED PRINCIPLE

One of the earlier principles that guided writers of lightning protection standards, the "cone of protection" theory, has long been discredited. This theory proposed that high objects tend to divert and intercept practically all lightning flashes that otherwise might have struck somewhere within a radius of one times the height of the object. Thus, an object 100 feet high above a surrounding flat area would tend to divert and become the target of practically all lightning flashes that might strike within a 100-foot radius or 200-foot diameter (see Fig. 9.3) and be the target of nearly all flashes that might hit within a 200-foot radius or 400-foot diameter.

"Cone of protection" proponents held that, in addition, the 100-foot object would tend to divert and intercept some small portion, say about 25 percent, of the flashes that would have hit outside the radius of two times the height of the object, but within a radius of four times the height.

Hypothesized in the nineteenth century, this theory has been shot full of "ifs" and "buts" in the course of time. Application of the theory, without taking into account the many exceptions, not only results in ineffective lightning protection systems but, as happened in the case of one unwitting architect, could cause serious legal and financial difficulty. The theory came to be known as unreliable for the following reasons.

At the first writing of the cone theory, churches and schools were compact buildings with relatively tall steeples and chimneys, respectively. Thus, architects and owners got into the habit of specifying "steeple systems" and "chimney systems." The practice was carried into other structures with tall sections.

Through the years, design preferences have shortened towers. Fuel changes and emission restrictions have altered smokestack configurations. Designers who cling to old practices sometimes do not even meet cone theory requirements, settling, for economy's sake, on partial protection. In one real-life example, a church building committee rejected a complete lightning protection system and settled for

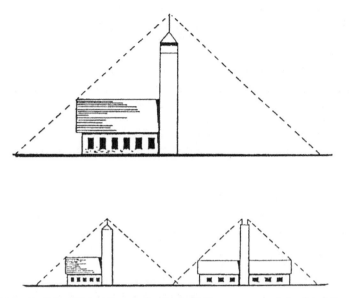

FIGURE 9.3 Until the 150-foot diameter "rolling ball" theory was accepted, a 1-to-1 "cone of protection" was used to determine the extent of lightning protection required.

a steeple system only. The church was struck at the far end and ignited even before construction was completed. The architect and builder were sued by the insurance company, with costly results.

In an even more dramatic incident, a storage structure used by the U.S. Navy for unstable propellent fluids had been protected by single grounded poles located near the building's four corners. The protection scheme conformed to code requirements of World War II vintage, when single grounded poles were assigned protective ranges extending outward at a 45 degree (1 to 2) angle from the tip of the lightning rod atop each pole. An explosion occurred coincidently with a thunderstorm in the area. The blast was so intense that the structure and contents disintegrated, and trees were scorched in a wide area around the site. While certain other causes of the explosion to the untended facility were contemplated, the consensus of members of the National Fire Protection Association's Lightning Protection Code Committee was that the lightning protection system failed to provide adequate coverage, and a lightning strike to the structure caused the blast.

ZONES OF PROTECTION

As a result of plotting the locations of lightning strike points over periods of time, atmospheric scientists and other analysts have showed that, for structures more than 50 feet tall, straight-line delineation of strike and nonstrike areas was incorrect. Instead, they proposed the "rolling sphere" theory now used in determining the protective coverage provided by tall structures to lower structures and surrounding ground areas.

In this technique, a sphere 150 feet in diameter is theoretically brought up to and rolled over the total building (see Fig. 9.4). All sections of the building that the sphere touches are considered to be outside the zone of capture capability that is provided by air terminals located on higher sections. Therefore, it is likely to receive direct strikes. Sections of the building that can not be touched by the rolling sphere are considered to be in a zone of protection and immune from lightning strikes.

For unusual or complex building forms, the "rolling ball" technique may be used directly in determining both zones of protection and air terminal configurations. Buildings that exceed 50 feet but do not exceed 150 feet in height are considered to protect lower sections or structures where these lie in the space beneath an arc of 150-foot radius, and where the arc passes through the highest point of the building and is tangent to ground.

Buildings more than 150 feet high are considered to protect only those lower sections of a structure that lie in the space beneath an arc of 150-foot radius that is tangential to both the side of the building and the ground (see Fig. 9.5).

If a building is more than 150 feet tall, direct strikes to the side of the structure above the 150-foot level may be anticipated. However, such side strikes have been shown to be inconsequential in electrical magnitude and, therefore, also in de-

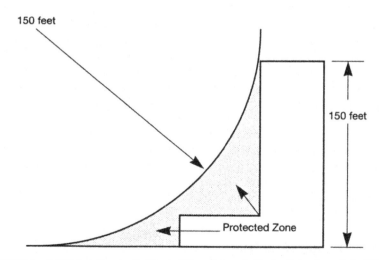

FIGURE 9.4 A 150-foot diameter "rolling ball" template scaled to the dimensions of a building can be theoretically rolled up to and over a building to determine coverage provided by higher, protected roofs.

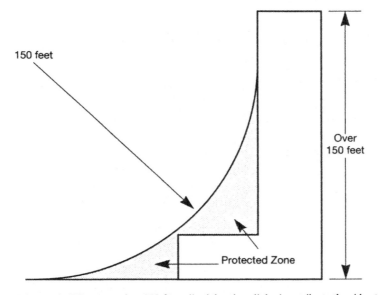

150 feet

Over
150 feet

Protected Zone

FIGURE 9.5 A building more than 150 feet tall might take a lightning strike to the side at a point above the 150-foot mark, but such strikes have been shown to be weak and not destructive to buildings.

structive ability in the case of buildings and other structures of ordinary construction and noncritical contents.

Because lightning has such a wide range of consequences, perfect protection under all conditions may be difficult to obtain without high cost. But the degree of protection achieved with proper design and installation, using proper equipment, is considered to be as high as 99.9 percent, given, of course, periodic inspection and any required maintenance.

PROTECTION FOR LOWER SECTIONS

Buildings that do not exceed 25 feet in height are considered to protect lower sections of the structure in a two-to-one zone of protection. As shown in Fig. 9.6a, the 45 degree, straight-line method of determining protection is easier to apply to low buildings than is a simulated 150-foot radius rolling ball. It adapts itself readily to ordinary residential-type construction, and it conforms to the angle provided by the lower 25 feet of height of a 150-foot radius "rolling ball."

Buildings taller than 25 feet, but no taller than 50 feet, are considered to protect lower sections of the structure in a one-to-one zone of protection angle as shown in Fig. 9.6b. If you draw a straight line on a 90 degree angle and lay it on a 150-

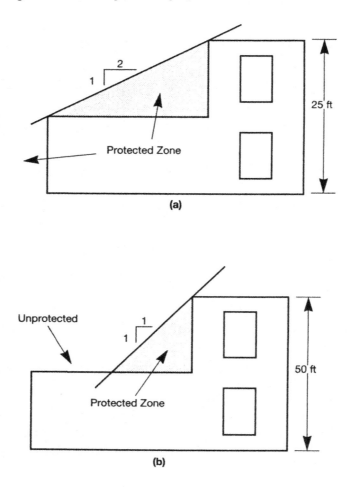

FIGURE 9.6 The lower roof of a building up to 25 feet tall (a) is protected by a higher section if the lower roof is within a 2-to-1 diagonal line struck as shown. A building from 25 to 50 feet tall (b) offers a 1-to-1 protection zone for a lower roof.

foot radius rolling ball template, you will find that it conforms closely to the arc of the template between its 25-foot and 50-foot elevations.

When a structure is taller than 50 feet, a 150-foot radius "rolling ball," sized to the same dimensions as an elevation view of the building, can be rolled against, up, and over all sides of the building to determine which lower roofs, dormers, and other projections need to be equipped with air terminals, and which are protected by the air terminals on higher roofs and projections (see Fig. 9.7).

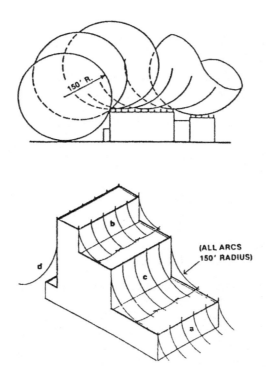

FIGURE 9.7 A simulated 150-foot diameter "rolling ball" can be rolled over a building to determine which lower roof areas of a building are protected by air terminals on higher roofs.

PROTECTION FOR DANGER STRUCTURES

Structures housing volatile or explosive materials may be subject to fire or explosion of the product or vapors by small-magnitude lightning flashes that strike within a 150-foot rolling sphere's zone of protection. For that reason, the radius of the sphere is reduced to 100 feet, as shown in Fig. 9.8.

Single-pole protection, as shown in Fig. 9.8, left-hand drawing, provides a limited area of protection because of the sag in the arc. Such protection is expensive, first because of its limited protective area, and because wood poles, which are reasonable in cost, are difficult to obtain in heights greater than 70 or 80 feet. Steel poles, which are comparatively expensive, can be obtained in greater lengths.

Overhead wire protection (see Fig. 9.8, right-hand drawing) is considered to be the most positive of all protection schemes. It is also inexpensive and can provide protection over a very large area at nominal cost. However, its use, in practice, is limited to those installations where esthetics are secondary to safety.

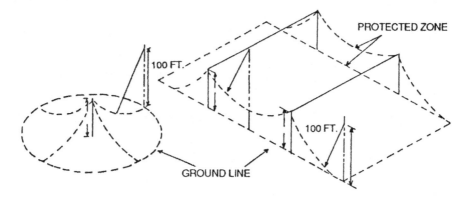

FIGURE 9.8 "Danger Structures," so named because of the unstable or volatile material they contain, can be protected by grounded poles or by such poles with overhead cables strung between them.

AIR TERMINAL CONFIGURATIONS

Traditionally, air terminals have been quite sharply pointed in conformance with Benjamin Franklin's early recommendations. However, Professor Emeritus Charles Moore of Langmuir Laboratories, Socorro, New Mexico, conducted an experiment that he reported proved the superiority of blunt rods over sharp. The experiment was conducted in a highly lightning-prone, mountainous area near Socorro, where atmospheric experiments are viewed and their results recorded in an underground observation chamber.

According to Professor Moore, the sharply pointed rod tended to protect itself by surrounding itself with an ion cloud that repelled downcoming stepped leader strokes, which then struck bushes and other nearby targets. The blunt rod, with its smoothly rounded tip, held a "bound charge" until the stepped leader neared, then released a powerful upward-moving, ground-based streamer, thus attracting the lightning current onto itself. As a result of Professor Moore's report, more and more blunt air terminals are being specified and installed.

AIR TERMINAL LOCATIONS

Air terminals are placed at locations that are likely to be struck by lightning and are spaced so it is considered impossible for a roof or other surface to be struck directly. On gable roofs, air terminals must be placed within 2 feet of gable ends, and at 20-foot intervals or less on roof peaks. On flat roofs, air terminals must be placed within 2 feet of roof edges and outside corners (see Fig. 9.9).

Air terminals that are 24 inches tall or taller may be placed at 25-foot intervals. Since air terminals that are up to and including 24 inches tall need not be braced, there is an advantage in using the 2-foot terminals at roof edges. However, owners

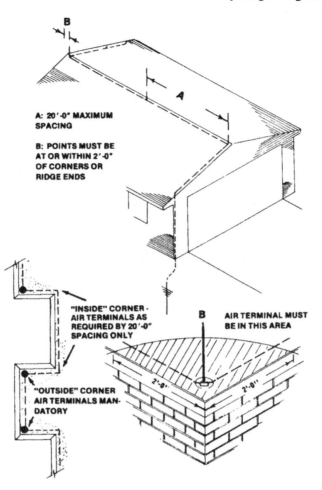

A: 20'-0" MAXIMUM
SPACING

B: POINTS MUST BE
AT OR WITHIN 2'-0"
OF CORNERS OR
RIDGE ENDS

"INSIDE" CORNER ·
AIR TERMINALS AS
REQUIRED BY 20'-0"
SPACING ONLY

"OUTSIDE" CORNER
AIR TERMINALS MAN-
DATORY

B AIR TERMINAL MUST
BE IN THIS AREA

2'-0" 2'-0"

FIGURE 9.9 Air terminals on roof peaks and edges normally are placed at 20-foot intervals and within 2 feet of roof ends, roof corners, and edges.

and architects often prefer that air terminals be as inconspicuous as possible. As a result, 10-inch units of 0.375-inch diameter are commonly specified.

Figure 9.9 shows an example where air terminals may need to be spaced closer than 20-foot intervals to comply with corner placement requirements. When air terminals that are taller than 24 inches are used, braces are available for top or side mounting as shown in Fig. 9.10.

The most likely locations for lightning strikes are roof corners, gable ends and edges, and tall rooftop structures. Since midroof areas of the main roof are least

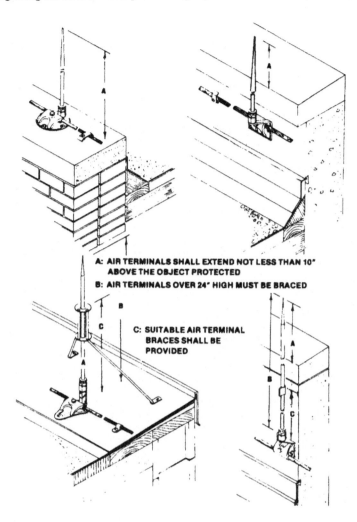

FIGURE 9.10   Air terminals must be equipped with suitable braces or to-wall supports when they are taller than 24 inches.

likely to be struck, midroof air terminals on flat or "gently sloping" roofs are permitted to be spaced at 50-foot intervals

CAPTURE FAILURES AND REMEDIES

During more than 30 years of involvement in the lightning protection industry, the author has been informed of four instances where air terminals failed to capture lightning flashes, which struck at varying distances from the terminals involved. In each such occurrence, the missed lightning rod was sharply pointed.

The most recent of such incident involved a vary large flat-roofed structure where, over the years, lightning sometimes failed to strike 12-inch tall air terminals located in midroof areas. Each errant strike was near the center of a grid formed by four midroof air terminals spaced at 50-foot intervals.

The region in which the capture failures occurred has frequent thunderstorms, and the large building is so situated that one side faces oncoming storms. Many air terminals on the building's roof exhibit melted tips, evidencing instances of successful performance by the lightning protection system. Fortunately, the building has thus far escaped fire or major damage during those instances when lightning flashes missed midroof air terminals.

There are two possible "fault and remedy" scenarios. In the first scenario, the fault could be insufficient air terminal height. Lightning protection standards call for air terminals at least 10 inches tall, wherever located. However, the Lightning Protection Institute's *Standard of Practice, LPI 175*, suggests the use of taller air terminals in midroof areas, with heights based on the 150-foot radius zone of protection arc. A table in LPI 175, based on the appropriate segment of such an arc, calls for 24-inch air terminals where the radius is 23 feet, and 36-inch tall air terminals where there is a protective radius of 29 feet. In scenario 1, then, the remedy is replacement of the 12-inch tall air terminals with 35-inch tall terminals.

In scenario 2, the fault is air terminal configuration, and the solution is a change in configuration, based on the experimental findings of Professor Charles Moore, as cited earlier. Under this scenario, the conclusion is that Benjamin Franklin, and the scores of engineers and lightning protection industry members who have served on the lightning protection code committee for the standard *NFPA 78*, have erred in regard to the most effective air terminal configuration.

In scenario 2, the air terminals should have blunt tips as recommended by Prof. Moore, Dr. Rodney Bent, and other atmospheric scientists. Sharply pointed lightning rods function properly most of the time. But, occasionally, they repel rather than attract, possibly because of factors related to the configuration (including time resolution) of the lightning flash involved.

It should be reemphasized that during the author's long involvement in lightning protection, several instances of failure by sharply pointed rods have been brought to his attention.

FLAT AND PITCHED ROOFS

Gently sloping roofs (see Figs. 9.11 and 9.12) are defined as roofs with

1. spans (dimension A) of 40 feet or less and pitches (dimension B) of less than 1/8, and
2. spans of more than 40 feet and pitches less than 1/4.

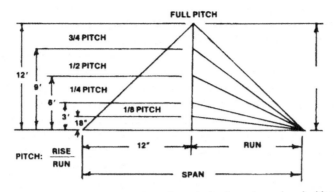

FIGURE 9.11 The pitch of a gable roof determines whether it can be equipped with air terminals on the peak only or must be protected at the edges.

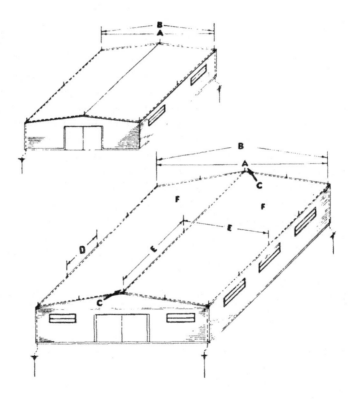

FIGURE 9.12 These gently sloping roofs are equipped with air terminals spaced at 20-foot intervals along roof edges and at 50-foot intervals in midroof areas—just as flat-roof buildings are treated.

(Shed type roofs are considered to be one-half of a a full-gable roof). Buildings with roofs that conform to those limits are treated in the same manner as buildings with flat roofs.

Air terminals are required to be spaced at 20-foot intervals around the perimeter of flat or gently sloping roofs if the air terminals do not exceed 24 inches in height. Where they do exceed 24 inches, air terminals may be spaced at 25-foot intervals around the perimeter.

The center areas of gently sloping roofs are required to have air terminals at 50-foot grid spacings. If "A" is more than 20 feet, an air terminal is required at each ridge end, regardless of the pitch or span (see Fig. 9.12).

PROTECTION OF ROOFTOP EQUIPMENT

Air handling units, vents, and other rooftop equipment can be utilized as mounting bases for air terminals. If the metal skins of such units are less than three-sixteenths of an inch (0.1875") thick, they are considered to be susceptible to "burn-through" by a lightning flash of prolonged duration and need to be protected with air terminals (see Fig. 9.13). When the skin of such a unit is more than three-sixteenths of an inch thick and the unit is properly bonded into the lightning conductor system, the unit may substitute for an air terminal.

Air terminals on residential-type chimneys (see Fig. 9.14) must be attached to the chimney so that no outside corner of the chimney is more than 24 inches from an air terminal. Air terminals, conductors, and other components must be hot dipped lead coated on the upper 24 inches of the chimney if made of copper. Aluminum air terminals need not be lead coated.

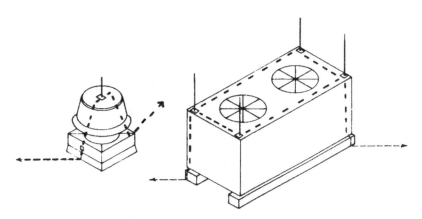

FIGURE 9.13 The skin thickness of rooftop air handling units and vents determines whether they need to be equipped with air terminals.

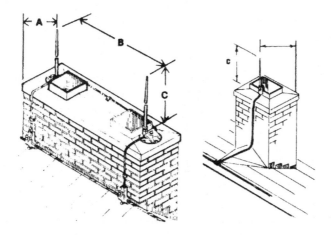

FIGURE 9.14   Air terminals on residential-type chimneys need to be lead coated if copper, but do not need such protection if made of aluminum.

Chimney air terminals must conform to standard distances and heights: A = 2 feet maximum, B = 20 feet maximum, and C = 10 inches minimum. Air terminals may be anchored directly to the chimney, as shown, or may be secured with a metal band around the chimney. During construction, it is possible to anchor the air terminals to the top of the masonry prior to pouring the concrete cap, thus concealing the chimney bases from sight and weather.

Copper air terminals on chimneys must be at least three-eighths of an inch in diameter. Aluminum points can be no less than one-half inch in diameter.

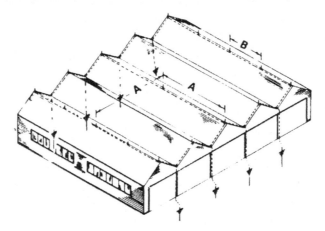

FIGURE 9.15   This multiple-ridge building is treated in the same manner as a flat-roof building in regard to air terminal placement. But the conductor system must provide two paths to ground for each air terminal.

DOMED AND MULTIPLE RIDGED ROOFS

Air terminals on buildings that have multiple ridged roofs (see Fig. 9.15) may be located and spaced in conformance with standard requirements. Air terminals on the outer gable roof sections (B) are spaced at 20-foot intervals, while air terminals on the inside roof sections (A) are spaced at 50-foot intervals. Note that down conductors and grounds may need to be placed unusually close together to provide two paths to ground from each air terminal. Air terminals on domed or curved roof structures must be located so that no part of the roof projects higher than the tips of the air terminals at that location (see Fig. 9.16).

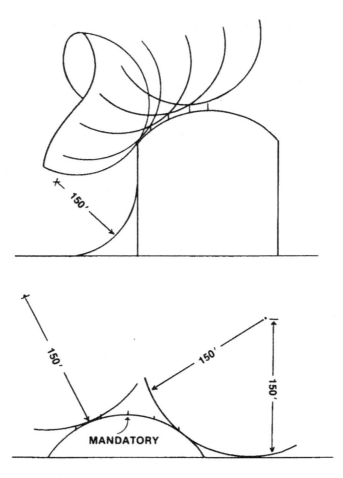

FIGURE 9.16 On domed or curved roof structures, care must be taken to assure that air terminals extend above the roof itself at all locations.

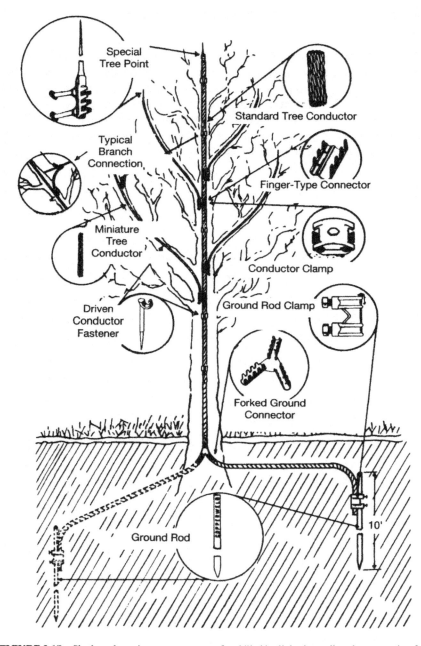

FIGURE 9.17 Shade and speciment trees are so often killed by lightning strikes that protection for them is common. This illustration shows proper protection of a typical spreading shade tree.

PROTECTION FOR TREES

Homeowners and owners of other types of low rise buildings sometimes rely on nearby trees to provide a form of lightning protection. It is the consensus of members of the code writing committee for the lightning protection code, *NFPA 78*, that trees do not provide protection from lightning to nearby buildings, and that a building provided with a lightning protection system does not extend that protection to nearby trees that are of equal or greater height.

The code writers recommend that trees with trunks located within 10 feet of a protected building, and branches that extend higher than the building, be individually protected. Newspaper clippings of lightning losses confirm that position. In many instances, lightning has struck trees and dug furrows to houses or other buildings to ground themselves in the lowest resistance earth in the vicinity.

Where trees are located near buildings, they may share grounding systems with those structures. When protected as individual targets, trees should be equipped with systems in conformance with their type and physical configuration.

A widely branched tree should be equipped with a standard tree air terminal at its highest available location, and secondary miniature air terminals on branches (see Fig. 9.17). A pine tree or a similar tree of narrow configuration usually requires only one main size air terminal located as high up as can be safely reached. Other species of trees may require one or two primary air terminals and as many as eight or nine miniature branch terminals.

Trees less than 3 feet in trunk diameter may require only one down conductor, while two down conductors are recommended for larger trees. Down conductors should be brought out from the trunk to ground rods or buried conductors located beyond the drip of the tree.

Trees may share grounding systems with buildings located within 25 feet and with other trees located no more than 80 feet apart, according to recommendations of the NFPA Lightning Protection Code Committee.

10

Conducting Lightning Current

When the stepped leader stroke of a large-magnitude lightning flash reaches striking distance above a building or other targeted structure, the stage is set for the transfer of a huge mass of electrical energy. The entire trip from thundercell to earth takes place in microseconds. This places a large burden on the conductors that must convey the charge from strike point to earth level and to the grounding electrodes.

Conductors must be used that offer minimal resistance to a large mass of high-potential electrical energy. This dictates that the conductors be of low-impedance material, make a minimum number of bends, and route the charge in as direct a path as can be achieved. As discussed in the preceding chapter, it is important to provide the current with at least two paths to ground, except in the case of small-magnitude flashes that may strike lower roofs.

CONDUCTOR TYPES, CLASSES, AND PURPOSES

All conductors used in lightning protection are of either copper or aluminum, and they are divided into three classes. These are designated Class I primary conductors, Class II primary conductors, and secondary conductors.

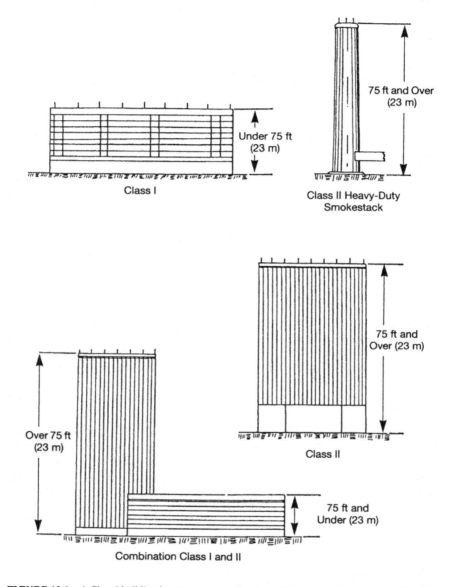

Class I

Class II Heavy-Duty
Smokestack

Combination Class I and II

FIGURE 10.1   A Class I building is a structure no taller than 75 feet. It may be protected with minimum conductor and air terminal sizes as shown in Tables 8.1 and 8.2 (Chapter 8). A Class II building is a structure taller than 75 feet, which requires components of larger dimensions than Class I structures (see Table 8.3). A building taller than 75 feet with a section lower than 75 feet may be protected with Class II materials on the high portion and Class I materials on the low part of the structure. A heavy-duty smokestack is one that is taller than 75 feet and has a flue size greater than 500 square inches. It requires Class II copper materials that are lead coated on the upper 25 feet of the stack.

Class I conductors may be used on any structure with a height of up to and including 75 feet. Class II standard conductors are required for taller structures (see Fig. 10.1). Secondary conductors may be used only to bond secondary metal bodies into the building's down-conductor system.

Secondary metal bodies are those within the protective zones of primary metal bodies and which are therefore not subject to direct strikes. Bonding, which is addressed in the following chapter, equalizes electrical potentials between grounded metal bodies, preventing the occurrence of a sideflash between them.

Class II lead-covered copper conductors (see Fig. 10.2) are used on the upper 25 feet of heavy-duty smokestacks. Copper conductors must be hot dipped, lead coated when used on the upper 2 feet of residential-type chimneys. Where aluminum conductors are used on chimneys, they need not be lead coated. However, aluminum conductors may not be used on heavy-duty smokestacks.

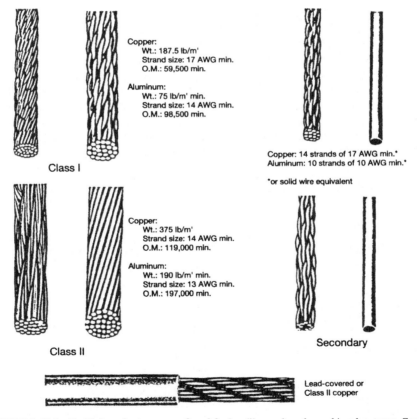

Copper:
  Wt.: 187.5 lb/m'
  Strand size: 17 AWG min.
  O.M.: 59,500 min.

Aluminum:
  Wt.: 75 lb/m' min.
  Strand size: 14 AWG min.
  O.M.: 98,500 min.

Class I

Copper: 14 strands of 17 AWG min.*
Aluminum: 10 strands of 10 AWG min.*

*or solid wire equivalent

Copper:
  Wt.: 375 lb/m'
  Strand size: 14 AWG min.
  O.M.: 119,000 min.

Aluminum:
  Wt.: 190 lb/m' min.
  Strand size: 13 AWG min.
  O.M.: 197,000 min.

Class II

Secondary

Lead-covered or Class II copper

FIGURE 10.2 Braided conductors are preferred for handling and workmanship advantages. Concentric conductors (as illustrated in the "aluminum" drawing under Class II) generally are limited to counterpoise grounding installations.

As identified in Fig. 10.2, Class I and Class II conductors are made up of several strands of small wires. Class II conductors must weigh at least 375 lb per 1,000 feet—without lead covering. Class I conductors must weigh at least half as much as Class II. Conductors used for equipotential bonding purposes only are reduced in size but cannot be smaller than a No. 6 AWG copper wire or its equivalent. They may be stranded or solid.

WHY AND HOW CONDUCTORS ARE STRANDED

Stranded conductors are required for current-carrying purposes in lightning protection partly because of the "skin effect" illustrated in Fig. 10.3. Electric current, whether produced by nature as lightning or by man as electricity, tends to flow on the outer surface of conductors where the least amount of resistance is encountered.

As a result of this skin effect, stranded cables offer less resistance to the passage of lightning's huge current flow than do solid conductors, and the latter would have to be larger in cross section.

Another reason why stranded conductors are the only ones permitted under United States standards is the fact that approximately one third of the members of the National Fire Protection Association's Lightning Protection Code Committee are representatives from the manufacturing and installation field, and another third of the committee are professionals—mostly electrical engineers. All are aware of the other advantages of stranded conductors: (1) ease of handling, (2) neater appearance, and (3) lower installation labor costs.

The Three Alternatives

Braided conductors rank first in order of preference among installing personnel, rope-lay conductors rank second, and concentric conductors rank far behind the

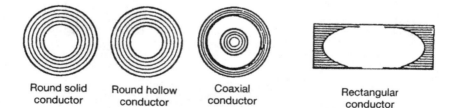

Round solid
conductor

Round hollow
conductor

Coaxial
conductor

Rectangular
conductor

FIGURE 10.3 Current Distribution in Conductors Because of Skin Effect. The "skin effect" that makes stranded conductors superior in conductivity is illustrated here. The heavier circular lines indicate concentrations of current flow along outer surfaces.

others. The reasons for the order of preference are based on ease of handling and esthetics. Braided conductors are the most pliable and therefore make for the neatest workmanship. They are also somewhat easier on ungloved hands than rope-lay cable. Concentric (twisted) cable, while the least costly by a large margin, is highly unpopular with installers because it is stiff and resistant to bending, and strands spring outward when the cable is cut.

Because of higher manufacturing costs, braided cable is somewhat more expensive than rope-lay cable and considerably more costly than concentric cable. Because of its low cost, the latter is widely used as buried counterpoise grounding conductors as well as other installations where long stretches of cable are laid in trenches.

LIMITATIONS ON USE OF CONDUCTORS

Copper conductors may be used underground as ground terminals where high bedrock precludes the use of driven ground rods. However, copper conductors, ground rods, and ground plates are subject to corrosion in soils that contain fertilizers, manure, or other caustic materials. Where such conditions exist, stainless steel electrodes should be used.

As mentioned, exhaust gases from chimneys and smokestacks are also corrosive to copper. For this reason, they must be lead coated on the upper surfaces of such structures.

Aluminum conductors need not be protected from smoke exhausts, but they are subject to corrosion where salt-laden air is present and where moist material may be in contact with them. Therefore, aluminum down conductors are permitted to extend only to within 2 feet of grade. Beyond this, a bimetal connector is used to convert the down conductor to copper cable.

Aluminum components should not be used in tree protection, where wet bark and moisture laden debris may be present. The mutually corrosive effects of copper and aluminum prohibit the use of copper components on aluminum roofing, and vice versa. Where copper conductors must pass over aluminum flashing, or aluminum over copper, and where copper or aluminum conductors must be attached to air handling units or other structures of the opposite material, lead or stainless steel buffers are required. Full-line lightning protection components manufacturers have various bimetal connectors available for different situations.

CONDUCTOR ATTACHMENT COMPONENTS
AND METHODS

Lightning protection standards require that conductor cables be attached at 3-foot intervals, and a variety of cable holders are available for that purpose (see

Fig. 10.4). Attachments to concrete and masonry walls are made by drilling holes and installing expansion anchors, which must withstand pull tests of 100 lb. Attachment of cables to shingled roofing on wood or composite sheathing is usually done with nail-on clips.

Two common attachment units are available for flat roofs. Where wood or other nailable decking exists under the roofing material, nail-on clips are used. Perforated cable holders are available for built-up tar and gravel roofing. The cable holders and conductors should be placed in position atop the tar or mastic and pressed down so that the mastic oozes up through the perforations in the holders

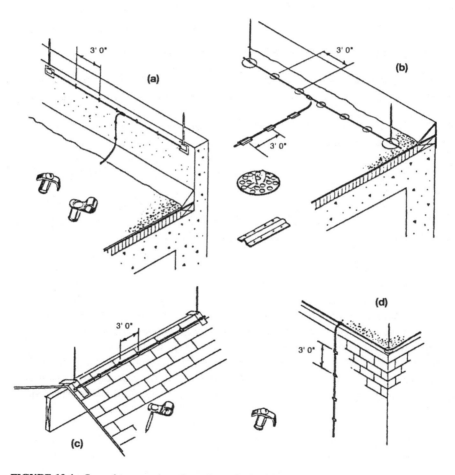

FIGURE 10.4 Several types and configurations of cable holders are available: (a) expansion anchors for concrete, (b) mastic anchored and nail or screw fastened cable holders for roof decks, (c) nail-on clips, and (d) expansion anchors for masonry mortar joints.

before the gravel is spread. Such a work sequence timing calls for communications between contractors, which the engineer may be able to facilitate. In retrofit installations on tar and gravel roofs, the gravel and tar may be scraped away, after which the holders are placed and the gravel replaced.

New monolithic roof coverings call for changes in lightning protection installation techniques and materials. Perforations must be kept to a minimum and require the close participation of the roofer. Where perforations are not permitted, the roof and rooftop structures should be carefully studied to determine if the latter can be utilized as mountings for air terminals. On retrofit installations, air terminals and cable holders may need to be mounted on blocks placed atop the roof covering.

CABLE SPLICER TYPES AND METHODS

Conductor cable splicers are available in four basic types and three sizes, as illustrated in Fig. 10.5. The different types are tee splicers, parallel splicers and cross splicers. The sizes are (1) main-cable-to-main-cable splicers, (2) main-cable-to-secondary-cable splicers, and (3) secondary-cable- to-secondary-cable splicers. Because they are subject to strong stresses as powerful lightning current travels through the conductors, cable splicers must be able to withstand pull tests of 200 lb.

ROOF-LEVEL CONDUCTOR ROUTINGS

Every lightning protection project requires consideration of roof-level conductor routings. On gable roofed buildings, particularly residences and other small structures, cable routing layout may be a simple matter with an obvious solution. But a flat roof building with parapets and rooftop air handling units, and with perhaps a penthouse or two and communications equipment, poses a more complex layout challenge. The routing question on a gable roof may simply rest on whether the structure is already in existence, under construction, or in the planning stage.

Roof conductors on a gable roof preferably are placed near the peak, where they can be fastened to a dual-purpose, saddle-type air terminal base, on the far side of the street or driveway entrance as illustrated in Fig. 10.6a. If the building is not yet roofed, the lightning protection installer has an opportunity to conceal conductors from view and weather by using a through-roof base as shown in Fig. 10.6b. Such semi-concealment of the lightning protection system may also be possible on an existing gable roofed structure that has an accessible unfinished attic. On flat roofs with roof-edge elevations, conductors can be made part of the air terminal as shown in Figs. 10.6c and d, thus eliminating the need for a second path from air terminal to rooftop.

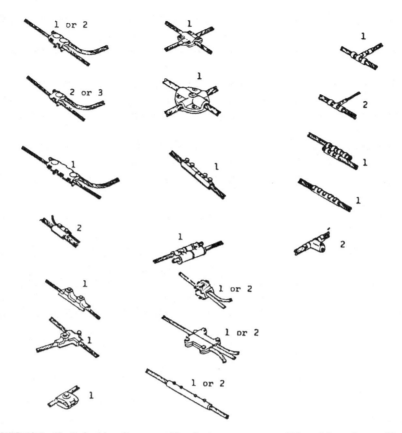

FIGURE 10.5 Typical cable splicers are of four basic types: tees, parallel, straight, and cross. Sizes are (1) main-cable-to-main-cable, (2) main-cable-to-secondary-cable, and (3) secondary-cable-to-secondary-cable.

A common variance from code requirements in installations by some installers is shown at left in Fig. 10.7. A "U" or "V" pocket causes a powerful resistance to lightning current, which may simply ignore the sharp reentrant loop and flash directly to the roof conductor. This creates a risk of ignition if any flammable material lies in a "hot bolt's" flashover path. Lightning protection standards do not specifically address the many individual questions that may arise with regard to conductor routings on roofs of varying edge configurations. Figure 10.8 illustrates four such instances. In each case, the main roof edge conductor runs parallel with the edge.

In Fig. 10.8a, a not-unusual wall configuration question is resolved by setting the air terminals and encircling conductor 2 feet back from roof extension corners. Where such corners extend farther than 2 feet from the main roof, one solution is

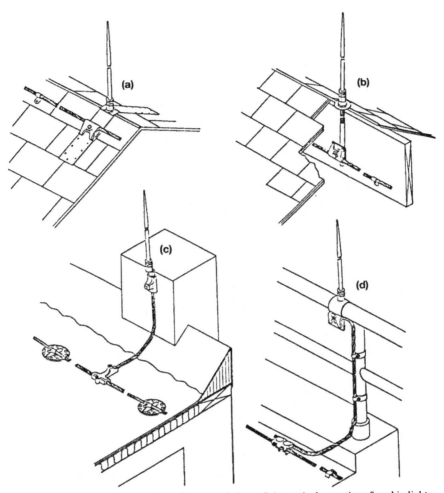

FIGURE 10.6 Shown above are four of many variations of air terminal mountings found in lightning protection: (a) a conventional saddle-type air terminal base and cable holder that bends to conform to the roof pitch, (b) a through-roof base preferred in new construction, (c) a permissible one-way-path installation, and (d) another installation that, in effect, makes the conductor a part of the air termination.

to use tall air terminals that provide a greater zone of protection. If that is not esthetically acceptable, it may be necessary to place air terminals within 2 feet of wall extension corners and to provide each with two paths to the main roof conductor.

In the roof configuration shown in Fig. 10.8b, the perimeter conductor rests on the main roof, where it connects with mid-roof conductors on the same level. In

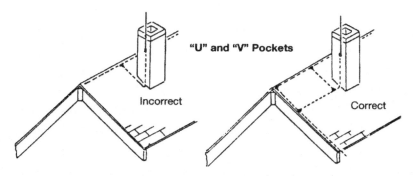

FIGURE 10.7 The "V" pocket in the left-hand drawing could be the cause of a "hot bolt" fire if lightning current were to short-cut the "V" and ignite flammable debris (e.g., dry leaves) at the chimney's base.

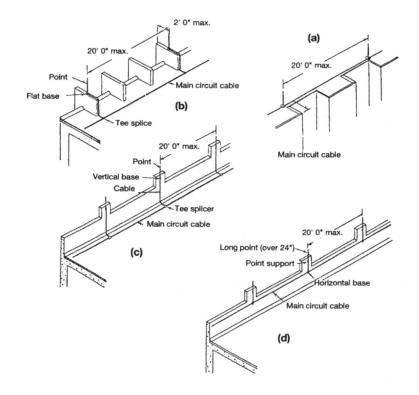

FIGURE 10.8 Many roof edge configurations confront lightning protection system designers and installers. Here, short runs of conductor from one-way paths to rooftop main conductors in drawings (b) and (c). Drawing (a) illustrates maximum spacings. Drawings (c) and (d) illustrate two ways of handling parapet installations.

these instances, individual conductor lengths are run up the parapet, where they are accepted as extensions of the air terminals. This is true also of the air terminals in Fig. 10.8c.

The roof configuration in Fig. 10.8c illustrates another common approach. The roof-edge conductor is run along the parapet wall at roof level, where conductors from mid-roof air terminals can run to the perimeter conductor at the same level.

The roof configuration of Fig. 10.8d shows still another common approach. The roof-edge conductor is run along the parapet wall, and air terminals are attached to it with appropriate bases. Other considerations aside, this installation method would be preferable to method of Fig. 10.8c when using copper conductors that must run over aluminum flashing.

Figure 10.9 illustrates two approaches to the problem of connecting rooftop air terminals to roof edge conductors atop or near the top of parapet walls. The top figure shows the preferred method. The conductor rises from the roof to the top of the parapet at the allowable rate of 1 foot of rise per 4 feet of length. Where conditions preclude rising along the parapet, the ramping method illustrated in the lower portion of the figure may be used.

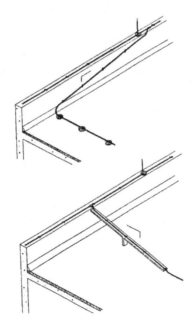

FIGURE 10.9 Abrupt conductor rises are limited by code to 8 inches. Where greater rises are necessary, they are limited to 1 foot of rise per 4 feet of length. Shown here are two permissible ways to effect such rises.

Another common problem encountered on larger existing flat roofed buildings is maintenance of a parallel path as a parapet mounted conductor encounters a roof edge elevation, such as an elevator housing. The problem lies in maintaining paths to ground from the air terminal on each side of the elevation.

If a down lead exists near the elevation, the ready solution is to drop down to the roof as shown in Fig. 10.10, method 1, and bring the conductor around the elevation at roof level. If there is no down lead in the immediate vicinity, method 2, which calls for drilling and sleeving through the structure's walls, can be applied. A third solution, not shown, is to carry a conductor around the elevation and maintain it at the same level as the parapet conductor.

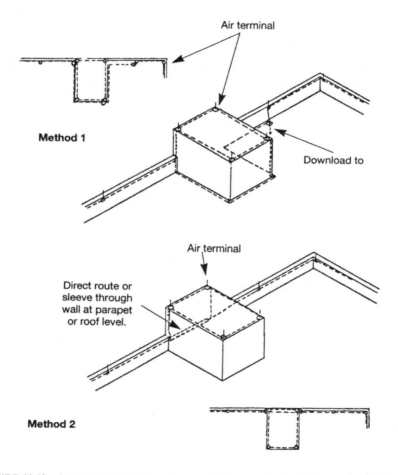

FIGURE 10.10 A common error likely to occur in lightning protection is failure to provide two paths to ground for each rooftop air terminal. Shown above are two solutions to such a problem.

PROTECTING OPEN AREAS IN FLAT ROOFS

Open areas in large flat roofs need roof edge protection, according to Code *NFPA 78*, " . . . if their perimeter exceeds 300 feet (92 meters), provided either rectangular dimension exceeds 50 feet (15 meters). Such protection, with appropriate air terminal and conductor locations, is shown in Fig. 10.11.

CONDUCTOR ROUTINGS AND BENDS

A common variation from lightning protection standards that is often visible in below-standard installations is a reentry loop at a conductor drop from roof level. In one extreme instance at a federal facility, the roof conductors on a flat-roofed building extended to the edges of a 4-foot wide overhang on each side of the structure, looped over the edge, and made reentry loops back to the exterior walls.

Whether lightning strikes had exerted outward thrusting forces on the conductors or the fasteners to the upper wall locations were substandard is not known. In any event, several of the conductors were loose and draped along the sides of the building. When high-pressure water is introduced into a hose lying loosely looped, the stream will tend to straighten the hose. The speed with which the straightening occurs and rigidity the hose achieves depend on the force of the water. Thus it also is with lightning current traveling along a loosely looped conductor. Firmly anchoring the conductor resolves the problem, but full assurance that it will remain resolved is achieved by limiting the bend to no more than 90 degrees. A reentrant loop in the conductor will produce an inductive voltage drop. The result may be a flashover across the open side of the loop. Disastrous results are possible, particularly if the lightning flash is a "hot bolt," and flammable material lies in the path of the flashover.

Lightning protection standards require that conductors make bends with radii no smaller than 8 inches, as illustrated in Fig. 10.12. The figure also illustrates another code violation, a "U" pocket created by incorrect conductor routing.

ONE-WAY DROPS AND DEAD ENDS

Figures 10.13 and 10.14 illustrate permissible dead ends and one-way drops of conductors. In the left-hand illustration of Fig. 10.13, a lower dormer is shown to have a total conductor length of 16 feet to an air terminal within 2 feet of its end, back to the main roof, and then across at the same level to a down conductor. The other two illustrations show that dead ends with one path to ground are permissible on wall sections as well as on dormers only when the total conductor length is no more that 8 feet.

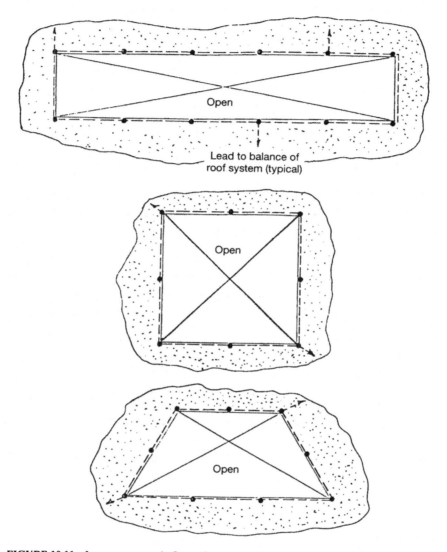

FIGURE 10.11 Large open areas in flat roofs are required to be treated as roof edges, with air terminals within 2 feet of their edges spaced at 20- or 25-foot intervals (depending on their heights) and two paths to ground from each air terminal.

Similar allowances are made for conductor drops from main roofs to lower roofs. In Fig. 10.14, a single path to ground is permitted, provided the conductor run at *A* does not exceed 40 feet. If location of the conductor drop to ground causes an interference, the conductor drop and ground terminal may be located at *X* or *Y*.

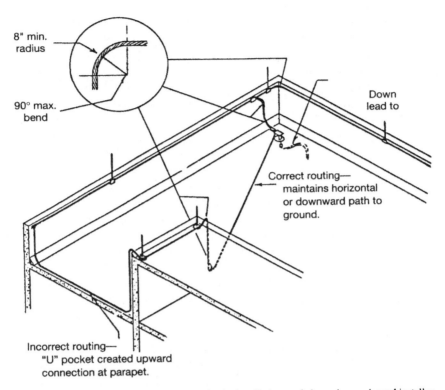

8" min.
radius

90° max.
bend

Down
lead to

Correct routing—
maintains horizontal
or downward path to
ground.

Incorrect routing—
"U" pocket created upward
connection at parapet.

FIGURE 10.12 Conductor Routing and Bends. An installation made by an inexperienced installer or one that shortcuts code requirements probably will have incorrect routings, creating "U" pockets and reentry loops.

THROUGH-ROOF AND THROUGH-WALL TRANSITIONS

Conductor transitions downward from the roof to a wall or an open space or through a wall can be readily accomplished using factory-made units that are adaptable to the dimensions of the roof or wall in question. Three of the most common through-roof transitions are illustrated in Fig. 10.15. Units are available that extend to an open space, such as a mechanical floor in a large building, or that simply extend through the roof covering and connect up with a conductor embedded in the wall.

Fig. 10.15 also touches on another important point, which is that aluminum conductors cannot be embedded in concrete or masonry where they might be subject to deterioration due to the presence of moisture. Where aluminum conductors are placed in walls, they are required to be run through plastic conduit. Copper conductors may be directly embedded.

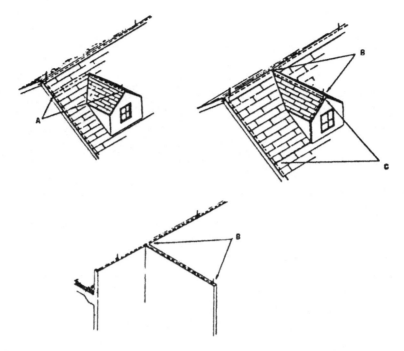

A: Permissible dead end—total conductor lengths not to exceed 16' 0"
B: Permissible dead end—total conductor length not to exceed 8' 0"
C: If "B" exceeds 8' 0", there must be a second path to ground as indicated.

FIGURE 10.13 Standard requirements with respect to "dead ends" apply to parapets and walls as well as to dormers.

One of the benefits of using steel through-roof units is that they can be used to effect a change from an aluminum roof system to copper conductors embedded directly in walls. Figure 10.16 illustrates several variations of through-wall connectors and methods. The threaded rods can simply be cut in order to fit tightly on walls of varying thicknesses.

CONNECTIONS TO AND FROM STEEL COLUMNS

Steel beams and columns with metal thicknesses of three-sixteenths of an inch (0.1875") or more are permitted under lightning protection standards to be utilized as primary conductors. This makes for a considerable saving in materials and labor, particularly in the case of a very tall building.

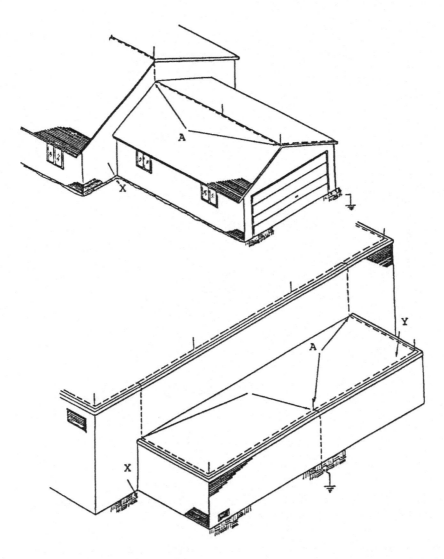

FIGURE 10.14 Experience has shown that lower roofs on low buildings are not struck by lightning flashes carrying high currents. Therefore, one-way conductor paths are permissible, provided that code requirements are met.

Individual conditions dictate whether it is best to bond air terminals individually to the steel framework at intervals conforming to spacing requirements, or whether roof conductors ought to be used. When a building is under construction, it may be possible to time air terminal installation so that direct connection to

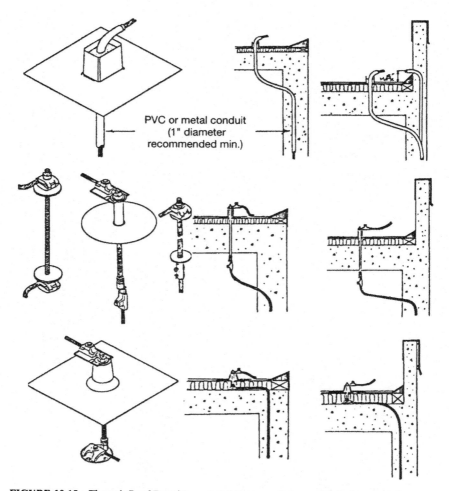

FIGURE 10.15 Through-Roof Transitions. A full-line manufacturer of lightning protection components can supply a variety of through-roof transitions.

beams or columns can be made. It is normally difficult and expensive to make such direct connections on an existing building, unless the work can be done coincidently with reroofing.

The left-hand illustration in Fig. 10.17 shows various types of connectors available to connect air terminals individually or via roof conductors to roof-level steel beams. The right-hand illustration shows common methods of attaching grounding connections to the bases of steel columns.

When air terminals are bonded directly to steel columns, their spacings must conform to standard requirements based on heights and locations. When roof con-

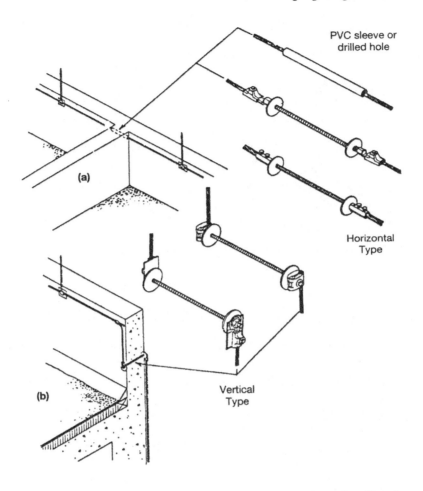

FIGURE 10.16 Factory-made through-wall connectors can save on-site time and trouble and ensure more positive water seals.

ductors are used, they are required to be bonded to beams or columns at intervals not exceeding 100 feet.

Rooftop air handling units and other metal bodies that do not come under the zones of protection of air terminals are required to be bonded to the building steel either directly or through roof conductors. Ground connections are required at every other perimeter column or at other locations around the perimeter where spacings do not exceed 60 feet. Air terminals, conductors, and connectors used on buildings utilizing steel framing in lieu of cable down conductors must conform to Class I or Class II size and weight requirements according to height.

Connections to steel
at roof level

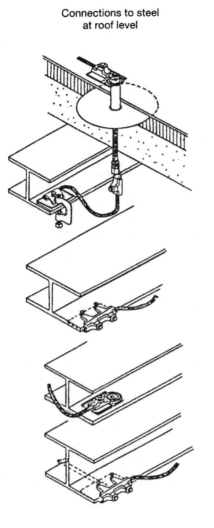

Connections to steel and
leads to ground at grade

FIGURE 10.17 A variety of connectors to and from steel columns can be employed to fit varying conditions. Exothermic welds are also permissible under code.

GUARDS AND DISCONNECTORS

Down conductors located in areas where people or animals might be present during thunderstorms are required to be guarded for safety as well as to prevent physical damage to the conductor. Figure 10.18a shows four types of such guards.

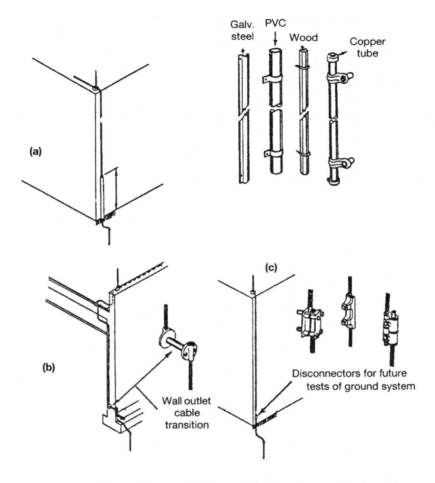

FIGURE 10.18 Conductor guards are available for people and animals, as well as the conductors themselves. Also available are wall outlet cable transitions and disconnectors for use when tests of the grounding system are desired.

Where horses, cattle or other large animals might be present to rub against a down conductor, it should be guarded with either a galvanized steel or copper tube guard. Keep in mind that an aluminum down conductor should not be placed inside a copper guard.

Where people are likely to be present, the guard should be nonconductive (either PVC or wood) to prevent a skin burn on someone who may be in contact with the bare conductor. For normal use, such protective guards are specified to be 6 feet high.

WALL OUTLETS AND DISCONNECTORS

Figure 10.18b shows a through-wall cable transition device that can be used to convert an aluminum conductor in a basement, or on an inside wall to copper, for grounding, or simply to bring a copper conductor to the outside.

Figure 10.18c shows alternate in-line clamps that can be used to disconnect down conductors at the base of a building so that a resistance test can be made of the grounding system.

11

Bonding to Prevent Sideflashes

The subject of bonding metal bodies to prevent sideflashing (also called "flashover") of current between them during lightning strikes has been a part of lightning protection code deliberations for generations. Early editions of the *Lightning Protection Code NFPA 78* specified that "metal bodies of conductance" and "metal bodies of inductance" be bonded into the lightning protection conductor system when they were found to be located within 6 feet of a lightning protection system conductor.

The terms and the "6-foot rule" appeared in code *NFPA 78* as late as the 1980 edition, as did a requirement that all metal bodies of conductance (which can be struck directly) were required to be bonded only if their area totalled 400 square inches or more. A more technical approach to bonding was sparked by the arrival on the committee of atmospheric scientist Dr. Rodney Bent, who wrote "An Explanation of Bonding Principles," which became the code's Appendix K.

STRUCTURAL BONDING

When a lightning flash strikes a tall, well grounded steel framed building that has been equipped with a properly installed roof system, most of the charge (scientists suggest 62 percent) will flow down the nearest column. The remainder of the current will be divided between the next two nearest columns until the top floor level is reached. There the current, having encountered a small amount of resistance to its downward travel, will be further divided.

This division of current will continue at each floor level until more columns accept a portion of the charge. Eventually, it is believed, all of the columns will carry a portion of the charge.

The chances are that there will be no sideflash of current to either interior or exterior grounded metal bodies as the charge travels toward earth. If there is, which might occur in the case of an extremely heavy lightning discharge, the sideflashes will be small ones—perhaps limited to inches.

On the other hand, when a lightning flash of similar magnitude strikes a tall, well grounded masonry or wood-framed building equipped with a similar, properly installed roof system, but where down conductors are spaced at maximum distances permitted by code, the results could be disastrous under some conditions, even though the charge has a shorter path to follow.

Masonry and wood-framed structures rarely rise as high as a dozen stories. Nevertheless, large flashovers may take place unless such a building is equipped with bonding conductors to equalize the electrical potential.

A reinforced concrete structure is betwixt and between the two types of buildings cited. Reinforcing bars in the concrete will substantially assist lightning's downward travel, particularly if the reinforcing bars are wire tied to each other.

WHY FLASHOVER OCCURS

Sideflashes are caused by differences in potential (voltages) between the current flowing along the conductors of a protected building and nearby metal bodies such as pipes, stacks, and wires. Such potential differences are the result of (1) resistance by conductors to the passage of electric current, and (2) magnetic field induction.

Copper conductors offer the least amount of resistance among the metals used in lightning protection. Aluminum offers 1.5 times the resistance of copper, and stainless steel 10 times the resistance of copper.

Because copper and aluminum are the only metals commonly used in lightning protection conductors, and because they are so close together in conductivity, no distinction is made between the two metals in determining bonding distances. The effects of magnetic field induction produced is also presumed to be the same whichever of the two metals is involved.

Current flashover, then, occurs when the combined resistive and inductive effects reach a point at which they overcome the resistance offered by the air and/or other nonconductive material existing between the conductor carrying the current and another grounded metal body.

In the first case cited, where lightning strikes a well grounded steel framed building, the resistive and inductive effects don't have a chance to cause any significant amount of resistance before the floor beams on either side of the first conducting column present themselves as cross conductors to alternate grounded columns.

In the second case cited, where lightning conductors are spaced at the maximum 100-foot intervals, resistive and inductive effects continue to build rapidly until they overcome the resistance of air between the load carrying conductor and the nearest grounded conductive body.

The conductive body involved might be a metal water pipe, a vent stack, an electrically powered machine, a computer, or (in the worst case) a person who is in contact with electrically powered equipment.

THE RESISTIVE EFFECT

Figure 11.1, which is a reproduction of Fig. K in the Appendix section of *NFPA 78*, illustrates both the resistive and inductive effects of lightning current flowing along conductor *A-B-C*, which is in close proximity to the separately grounded water pipe at *C-D*. In this example, the author of the section, Dr. Bent, assumed a resistance of 20 ohms (Ω) between *C* and ground, and a worst-case current flow of 100,000 amperes (A). Using Ohm's Law, which holds that voltage is equal to the current times the resistance, the potential on the conductor *A-B-C* is 2 million volts (V).

Since no current has yet been introduced into it, water pipe *F-E-D* is considered to be at zero voltage. The 2 million volt potential difference between the conductor and water pipe at *B-F* is powerful enough to produce a sideflash of more than six feet. A bonding conductor, *C-D,* installed at or near ground level, will equalize the potentials and remove what, in this case, is the inevitability of a sideflash.

LIGHTNING'S INDUCTIVE EFFECT

It is helpful at this point to repeat the earlier description of the causes and possible consequences of lightning's electrical effect:

When a heavy flow of rapidly varying current is introduced into a lightning conductor, the inductance of the conductor may combine with a strong resistance at ground level and sufficient current flow duration to produce a spark to another nearby grounded object. This "sideflash" is the result of lightning's need to establish additional paths on which to unload high induced voltages that exist for about one microsecond at the beginning of each return stroke.

Flashover frequency and distance are dependent on building height (which determines the length of the conductor and therefore the amount of voltage induced into it), the number of conductors present to carry current downward, and the amount of resistance encountered at ground level. Since those factors influence flashover, prevention of such sparking calls for measures to decrease induction, increase conductivity, and lower ground resistance."

The rapidly varying current referred to produces a magnetic field in a circular motion. The lines of magnetic flux intercept the loop *B-C-D-E-F*, inducing a voltage in the loop that is different from the potential existing in the down conductor *A-B-C*. This induction produces several million volts of potential difference between *B* and *F*, which will be relieved by a sideflash. The solution, again, is bonding at *C-D* and perhaps at *B-F*, depending on length of the pipe at *E-D*.

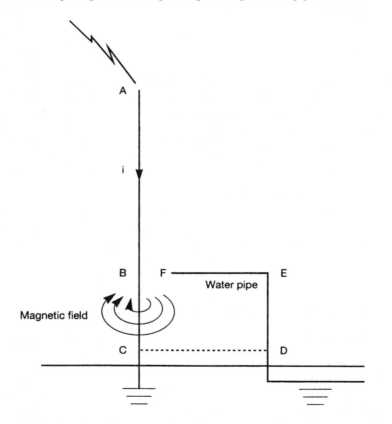

FIGURE 11.1 A resistance to current flow from *A* to *C* combines with magnetic induction to build up a potential difference as high as 2 million volts in this illustration from the National Fire Protection Association's *Lightning Protection Code NFPA 78*. That amount of voltage is sufficient to cause a flashover of more than 6 feet at *B-F*. The danger of such a sideflash is eliminated by equipotential bonding at *C-D*.

THE BENDING DETERMINATION FORMULA

The difference in electrical potential produced at the gap *B-F* depends primarily on the height of the pipe, *D-E,* and to a much lesser extent on the horizontal length at *C-D.* The latter is ignored, and the formula used to determine at what distance (D) bonding is required is:

$$D = \frac{h}{6n} \tag{7.1}$$

where,

  h = the vertical distance between the nearest existing bond (which may be at roof level, at ground level or at an intermediate point) and the bond being considered

  n = number of down conductors located no more than 100 feet from the bond being considered

FACTORS AFFECTING THE NEED FOR BONDING

As discussed previously, tall buildings sometimes act as "triggers" for lightning by bringing ground potentials, which are usually positive, up to the top of the structure. This shortens the distance a negative stepped leader stroke must traverse to reach effective ground. Tall buildings also sometimes initiate large "positive strikes" where no negative stepped leader stroke precedes the upward flow of current.

Subject though they are to "triggered" lightning in addition to "normal" lightning, very tall buildings do not usually require bonding beyond that which is inherent in their construction. However, they should still be examined to see if any long, vertical interior metal bodies exist that are not inherently bonded at 60-foot intervals or less.

REDUCING THE NEED FOR BONDING

Building owners and managers sometimes prefer to have down conductors run along stairwells or through other available interior spaces for esthetic reasons. When that is the case, it should be understood that costs may be substantially higher for a number of reasons, including the need for more extensive bonding.

The intensity of lightning's electric field in air is approximately twice that of an identical field encased in a dense material. Therefore, the bonding distance determined by the formula can be divided by a factor of two when down conductors are run on outside wall surfaces rather than inside the walls.

The need for bonding can also be reduced by increasing the number of down conductors, spacing them closer together, and by providing encircling conductors at intermediate elevations to simulate the effect of steel framing.

ROOF-LEVEL POTENTIAL EQUALIZATION

When lightning strikes a rooftop object, a large resistance exists as the current seeks to make its attachment to the object. If the struck object is not grounded, flashover will take place to the nearest grounded object or objects, regardless of the amount of electrical potential. On the other hand, if the struck object is grounded, the possibility of sideflashing depends on the electrical magnitude of the flash and the distance to other grounded objects. Figure 11.2 illustrates bonding requirements for primary and secondary objects on a gable roof. Included are one primary object, three secondary objects, and one object that need not have been bonded.

The latter object is (1) an ungrounded roof vent bonded with a crimp type fastener. The primary object (2) is the main plumbing stack, bonded with a bolt-compression stack band and provided with two paths to the main roof conductor. The three secondary metal objects are (3) a grounded sheet metal vent stack bonded with a crimped metal stack band, (4) a ventilator with a bonding plate fastened with sheet metal screws, and (5) a gutter bonded with a crimp-type gutter connector.

Rooftop television antennas (see Fig. 11.3) are required to be bonded to the conductor system with a full-size cable. In this instance, the antenna is adjacent to the main roof conductor and therefore requires only one path to that conductor. An arrester is attached to the TV lead-in wire.

Figure 11.4 illustrates six different rooftop objects bonded to the conductor system of a flat-roofed building with a penthouse atop it. Each of the bonded objects is attached with a different type of connector.

Lightning Protection Code NFPA 78 requires that grounded metal bodies in or on a structure taller than 60 feet be interconnected within 12 feet of the main roof level. On flat or gently sloping roofs, such bonding is best achieved by attaching to the roof perimeter conductor as illustrated in Fig. 11.4.

In the case of a tall, pitched-roof structure, *Code NFPA 78* requires that a conductor loop be placed at eave level, and that primary and secondary metal objects be bonded to that conductor.

It is important to keep in mind that two paths to the encircling conductor are required for primary metal bodies, and only one is required for secondary metal bodies.

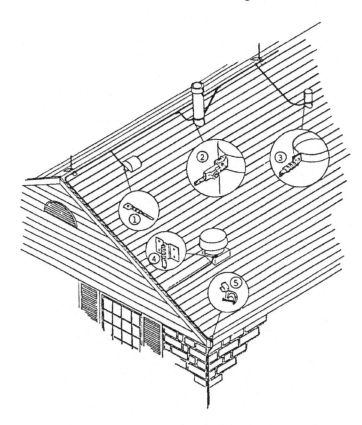

FIGURE 11.2 Low gable roof structures such as residences pose relatively simple bonding challenges. In this illustration, stacks 2 and 3, vent 4, and gutter 5 require bonding because they are grounded. Vent 1 did not have to be bonded because it is simply an ungrounded attic vent. The gutter is grounded via a downspout extending to grade.

GROUND-LEVEL POTENTIAL EQUALIZATION

All grounded metal bodies are required to be bonded to the lightning protection system at or near ground level. For residences and other low buildings 60 feet tall or less, bonding conductors may be connected to the nearest down conductor as illustrated in Fig. 11.5. The down conductor utilized may terminate at an individual ground terminal that meets grounding requirements as determined by soil conditions. However, any building that is more than 60 feet in height is required by *Code NFPA 78* to be equipped with a counterpoise grounding electrode, which is an encircling conductor installed underground and in contact with conductive earth.

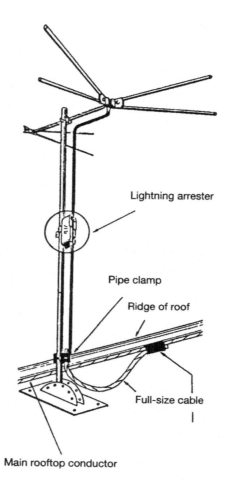

Lightning arrester

Pipe clamp

Ridge of roof

Full-size cable

Main rooftop conductor

FIGURE 11.3 Rooftop television antennas ranked second as lightning strike points in a statistical study involving residential lightning-caused losses. This antenna has been made into a functional air terminal that serves as a potential capture point for lightning.

INTERMEDIATE-LEVEL POTENTIAL EQUALIZATION

As discussed earlier, a steel-framed building provides potential equalization at each floor level when the framing in electrically continuous. Therefore, no further bonding is required.

A reinforced concrete structure that is more than 60 feet high also possesses some roof-to-ground electrical continuity, provided the reinforcing steel is inter-

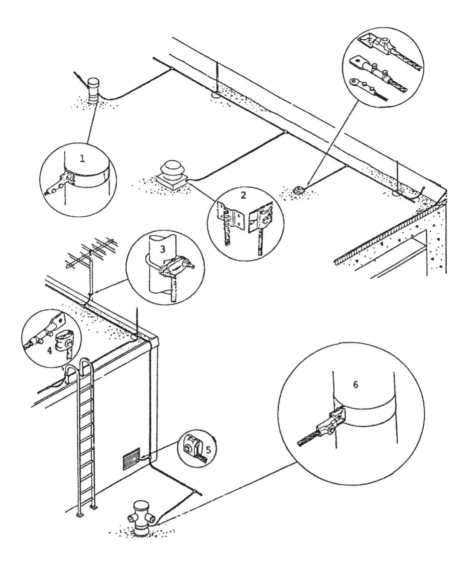

FIGURE 11.4   Among these six rooftop bodies are four secondary grounded metal bodies and two primary metal bodies. The former are considered to be immune from direct lightning strikes due to the protective zones offered by air terminals. Therefore, they need be tied only to the primary lightning conductor with a one-way path secondary conductor.

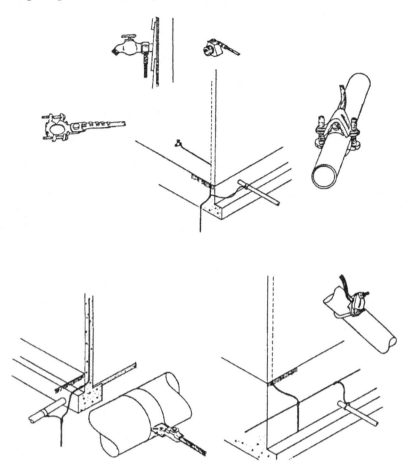

FIGURE 11.5 All utility lines entering and exiting a building are required to be bonded to the light-ning protection system. In the case of a structure less than 60 feet in height, utility lines may be bonded to individual down conductors as illustrated. For taller structures, utility lines must be bonded to a counterpoise conductor.

connected and grounded according to building code. However, a certain amount of impedance which could lead to sideflashing during a high magnitude strike is considered to be present. Therefore, the lightning protection down conductors and other grounded metal bodies are required to be interconnected with an encircling conductor at 200-foot intervals or less.

Masonry wall and wood-framed structures require interconnection of down conductors and other grounded metal bodies at intermediate intervals of 60 feet or less. Again, intermediate connections must be in the form of loop conductors.

BONDING CONDUCTORS

Conductors used to interconnect down conductors at the intermediate levels cited are required by code to be of the same weight as the down conductors themselves. Conductors used to equalize potentials between grounded metal bodies and metal bodies within a zone of protection may be of secondary size, as illustrated in the preceding chapter.

Bonding Interior Metal Bodies

NFPA 78 specifies that lightning should be permitted to travel no more than 60 feet before down conductors are bonded to prevent sideflashing. That restriction applies to interior metal bodies such as pipes and stacks as well as lightning down conductors.

It is true that most interior metal bodies are inherently bonded by virtue of connection to metal plumbing or electrical lines. But there are instances where that is not the case. In such instances, grounded and ungrounded metal bodies that exceed 60 feet in vertical length are required to be bonded as close to their extremities as is practical.

In the case of a structural steel building, the bonds should be framing members. In the case of a reinforced concrete building, the bonds should be to the lightning conductors. If the building is wood framed or has masonry walls, bonding locations must be determined by a formula.

When grounded metal bodies are found to be connected to the lightning protection system at only one point (at roof level or ground level), the following formula should be used:

$$D = \frac{h}{6n} \times k_m \tag{7.2}$$

where,

$h =$ the vertical distance between the nearest existing bond and the bond being considered

$n =$ the number of down conductors located within 100 feet of the bond under consideration (with 6 being a control factor)

The factor k_m is equal to 1.0 if the flashover is through air, but it is reduced to 0.5 if the flashover distance is limited because the current must pass through concrete, brick, wood, or other dense material.

The above formula should be used wherever the bond in question is located. However, factor "n" varies according to location. When the required bond is located within 60 feet of the top of the structure, "n" refers to the number of down-conductors that are spaced at least 25 feet apart and are located within 100 feet of the bond under consideration. When the required bond is located below the upper 60 feet of the structure, "n" refers to all of the down conductors in the lightning protection system.

Bonding Isolated Metal Bodies

Nongrounded metal bodies such as window frames should be taken into account when designing a lightning protection system because they may act as current-conducting bridges between grounded conductors. Figure 11.6 illustrates how a metal window frame can shorten flashover distance and how potential equalization can be affected by bonding, thereby eliminating the possibility of a flashover occurring.

How to Reduce or Eliminate Bonding

The best time to eliminate or minimize the need for bonding is during the system design stage. That's easy enough in the case of a new building, but it requires an

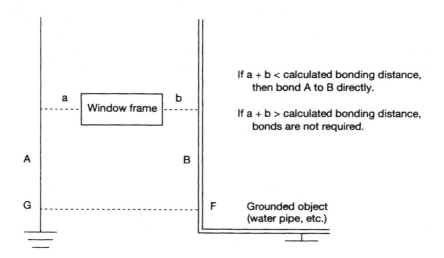

FIGURE 11.6 A metal window frame may serve as a conductive bridge for a flashover between a lightning conductor *(A-G)* and a grounded object *(B-F)*. To eliminate the possibility of a sideflash *(a-b)*, equipotential bonding must be effected. In this case, a bonding conductor from *A-G* to the window frame, and from the window frame to *B-F,* would prevent the occurrence of a sideflash. Where there are several closely spaced window frames, it is advantageous to bridge from *A-G* to *B-F* with a single bonding conductor.

on-site survey—possibly an exhaustive one—when you are working with an existing structure. In either case, each of the four bonding reduction measures cited below should be considered.

1. First, the code-dictated bonding requirements for all grounded metal bodies and those nongrounded metal bodies that form bridges between grounded metal bodies should be fulfilled.
2. Conductors should be placed within exterior walls in every new construction project, for the sake of reducing bonding distances by half as well as to visually conceal conductors. In existing construction, building owners should be made aware of the cost reduction benefits that may be inherent in placing conductors on the outside of walls rather than on inside surfaces.
3. Ground resistance should be minimized as much as possible.
4. Finally, serious thought should be given to the advantages of employing more down conductors so that more than one such conductor can be incorporated into the bonding strategy.

Examples of the reductions that can be achieved in flashover probability as well as flashover magnitude are illustrated in Fig. 11.7, parts a through d. Assume that the masonry wall structure houses many metal bodies of inductance. Assume also that these bodies are grounded by metal pipes that rise from the floor in such a way that no bonds exist in the lightning protection system except on the roof and at the ground floor.

In Fig. 11.7a, down conductors are run on the inside surfaces of the outside walls and are spaced at the maximum distance of 100 feet. Using Eq. (7.2), we compute:

$$D = \frac{60}{6 \times 1} \times 1 = 10$$

Therefore, we need to bond any grounded metal body that is situated within 10 feet of a conductor to that conductor.

Obviously, then, running down conductors along the interior surfaces of exterior walls, in stairwells or other interior locations is not a good idea where sideflash currents can cause damage to sensitive systems and equipment.

In Fig. 11.7b, down conductors are placed on the outside surfaces of exterior walls, producing the equation:

$$D = \frac{60}{6 \times 1} \times 0.5 = 5$$

In this case, the bonding distance is 5 feet.

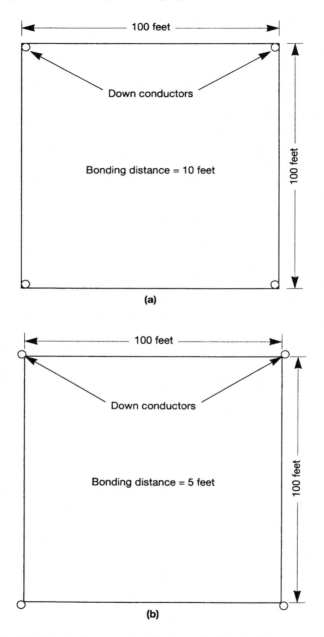

FIGURE 11.7 A building that houses many isolated metal bodies should be examined carefully to weigh the possibility of sideflashes against time, trouble, and cost of providing more down conductors.

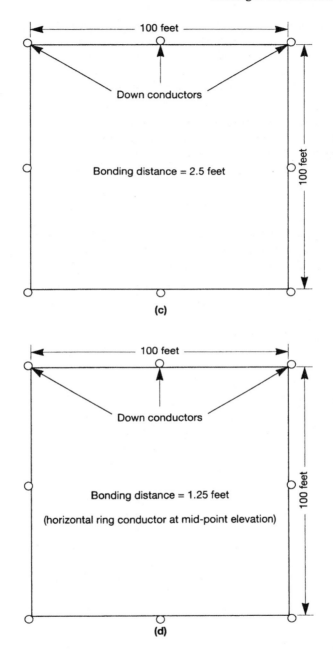

FIGURE 11.7 (continued)

In Fig. 11.7c, down conductors have been placed at 50-foot intervals. Here we compute:

$$D = \frac{60}{6 \times 2} \times 0.5 = 2.5$$

Finally, in Fig. 11.7d, an additional ring conductor has been installed on the building in question, reducing the vertical distance in the formula by half, to 30 feet. The formula then would read:

$$D = \frac{30}{6 \times 2} \times 0.5 = 1.25$$

Admittedly, the formula is based on worst-case lightning flash magnitudes, but it should be kept in mind that going beyond minimum in structural protection can be a prudent step where a building houses a multiplicity of costly and vulnerable high-tech equipment.

12

Lightning-Induced Surges

The technology described in Chapter 11, "Bonding to Prevent Sideflashes," was described as " . . . a part of lightning protection for generations." It was pointed out such bonding equalizes electrical potentials between different conductors or conducting surfaces so that no sideflash occurs between them.

In a sense, the goal of *surge suppression* is the opposite of that of bonding: surge suppressors are designed to prevent the rapid equalization of high voltage potentials during the occurrence of a transitory current surge. Such surges, commonly referred to as *transients,* are caused by the electromagnetic effects of lightning or by electrical switching. Surge suppression devices act to divert surges to ground (e.g., gas tube devices) or clamp transient voltages to an acceptable level (e.g., metal-oxide varistors and silicon avalanche devices).

SOURCES OF DAMAGING SURGES

There was a time in the not too distant past when lightning-induced transients posed no problem—when only high-level waves of excess current were of concern. But the proliferation of high-tech electronic equipment that is sensitive to relatively tiny current surges has created a separate new lightning protection technology. This generally is referred to as surge suppression.

Several phenomena produce high-voltage transients that can affect sensitive electronic equipment. Most are lightning-related, but some are discussed herein because their effects and remedies are similar to those of lightning.

1. A direct strike to a building that houses sensitive equipment
2. A direct flash to a powerline entering the building via overhead or buried wires
3. A direct flash to interconnecting wiring
4. A direct flash to a structure, tree, or other object
5. An intercloud or cloud-to-cloud flash in the vicinity
6. The presence of a charged cloud overhead
7. An electrostatic potential in surrounding air
8. Voltages induced on interconnecting lines
9. Transient currents caused by switching activity

Effects of a Direct Strike

A lightning flash to a building will instantly raise its electrical wiring system and metal frame by several thousand volts potential. Computers, control systems, and other sensitive equipment grounded to an electrode system isolated from the building system will remain at a potential different from that of the frame and power system ground. A potential difference of some thousands of volts will exist that can cause sensitive devices to fail.

However, if the computer and building grounds are interconnected, their potentials will be raised simultaneously, and no sensitive device will be exposed to overvoltage damage.

Effects of a Strike to a Powerline

A strike to a powerline that feeds a building will have the same effect as a direct strike to the structure, but damage is almost certain to be less severe (see Fig. 12.1). The surge suppressor at the building's powerline entrance will intercept much of the overage current and shunt it to the power company's grounding electrode. Interconnection of the computer and building ground wires at a single location will eliminate a voltage buildup between these units.

Effects of a Nearby Strike

When a nearby building, tree, or other object in the immediate vicinity is struck, electric energy will be induced into interconnecting wires. Quite often, the energy

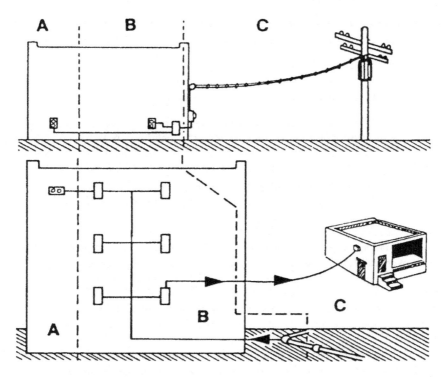

FIGURE 12.1 Exposure to current surge amplitude varies with location. In location A, exposure is limited by building wiring. At B, medium exposure occurs because the equipment is closer to the electrical service and building wiring provides less protection. In location C, high exposure causes major surge voltages and currents.   Courtesy of General Semiconductor Corp.

induced is of great enough magnitude to damage sensitive devices. To prevent such damage, surge suppression devices should be located on both ends of all interconnecting conductors.

Effects of a Cloud-to-Cloud Flash

The most common cause of damage to unprotected or improperly protected sensitive equipment is surges imposed on interconnecting wires by cloud-to-cloud or intercloud lightning flashes in the vicinity. The inductive consequences and the remedy are the same as for nearby flashes.

Effects of a Charged Cloud Overhead

When a thundercloud appears overhead, it brings with it a ground-based positive charge that in some cases may be of considerable magnitude. If sensitive equipment is linked to ground at a location distant from its main or parent unit, a voltage difference of several hundred volts can be developed in their vicinities, causing failure.

Effects of Atmospheric Electrostatic Potential

Even in fair weather, a potential difference exists. It is cited as being of a magnitude ranging from 100 to 600 volts per meter, depending on weather and height above sea level. During stormy weather, this differential may rise dramatically. Wires suspended above ground may pick up a portion of this voltage, causing an overvoltage and consequent equipment damage.

Overhead wires are susceptible to picking up of a portion of this voltage. Induced currents will be impressed on downline terminal components, often causing damage due to excessive voltage.

Such damage can be prevented by installing electrostatic shielding on all outdoor overhead lines. It is also possible to eliminate electromagnetically induced signals through use of surge protective devices at line terminals to prevent this type of overvoltage. However, continuous exposure of these devices to an aerial energy source can subject them to undue stress and lead to their eventual failure. For that reason, overhead signal wires should always be provided with grounded shields.

Environmentally generated charges can also result in the familiar electrostatic discharge (ESD) event, which most familiarly involves an accidental discharge of a transient into a sensitive circuit by a human being (see Fig. 12.2 and Table 12.1). ESD is similar to a lightning strike, but currents are small and rise times are very

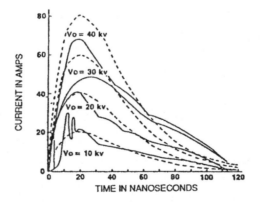

FIGURE 12.2 While lightning currents rise to their crests in microseconds, ESD charges typically do so in a few nanoseconds. This difference between millionths and billionths of a second requires surge protective devices to react much faster. Courtesy of General Semiconductor Corp.

TABLE 12.1   Electrostatic Discharge Voltages

Source	Highest	Most Common
Person walking across carpet	39,000 V	12,000 V
Person walking across tile floor	13,000 V	4,000 V
Person working at bench	3,000 V	500 V
16-lead ICs in plastic box	12,000 V	3,500 V
16-lead ICs in plastic tube	3,000 V	500 V

(Measurements taken at relative humidity of 15 to 36%)

fast. Table 12.1 (courtesy of General Semiconductor Corp.) shows actual voltages measured during several common activities. These range from a walk across a carpeted floor to handling integrated circuits. In the ESD field, primary emphasis is on prevention of the discharge rather than reactive measures.

Signal Effects

Nearby direct flashes, cloud-to-cloud flashes, and overhead cloud charges can cause induced voltages to flow into interconnecting lines. Such voltages may jumble normal signals by adding digits or neutralizing desired signals. A preventive measure is to run interconnecting wires inside rigid steel conduits that are grounded to the building's steel frame. While providing electromagnetic shielding from interference factors, this also eliminates the need for surge suppression devices on the wires.

Switching Transients

Transients caused by switching activity are closely related to lightning effects. As indicated by their designation, these transients are usually caused by switching functions in the electrical system. Such transients may appear as loads are switched on and off, or they may be generated outside the building in the power company's network (see Figs. 12.3 and 12.4). In certain cases, switching transients that are generated by one electrical utility customer have been known to cause problems for others tied to the same electrical system. Motors, transformers, and certain types of inductive heating equipment are notorious for producing switching transients.

We can gain an understanding of the mechanism by which a switching transient is generated by examining the operation of a simple direct current solenoid. Where a simple solenoid is connected to ground at one end, power is supplied to it through a switch. When the switch is closed, current will flow through the solenoid, establishing an electromagnetic field. In an electromechanical relay, this

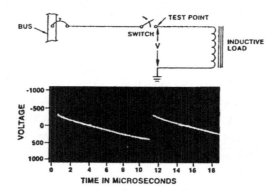

FIGURE 12.3 This figure illustrates that when power is supplied by closing the switch, current flow will establish an electromagnetic field that will collapse when the switch is opened. Courtesy of General Semiconductor Corp.

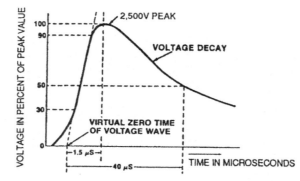

FIGURE 12.4 The waveform of a switching transient is quite similar to that of a typical lightning waveform. Because of the similarities, methods of protecting equipment against switching transients are often the same as those used for lightning. Switching activity, however, is normally higher. Courtesy of General Semiconductor Corp.

field is used to operate the armature. The strength of the field will depend on such factors as the number of turns in the solenoid winding, the type of core material used, and the power applied.

When the switch is opened, there no long is an applied electromotive force to support the magnetic field in the solenoid, and the field will collapse on itself. This time, however, instead of the current in the winding producing a magnetic field, the magnetic field produces a current in the windings. Because the energy stored in the magnetic field cannot be destroyed, it is returned to the circuit as electricity.

The voltage appearing at the open switch will be opposite in polarity to the originally applied potential and may be up to 10 times higher. The amplitude of the voltage will also depend on the speed at which the switch is opened. For a simple switch, this opening may be on the order of a few thousandths of a second. When solid-state switching is used, the switching speed may be several times faster.

A high difference in voltage will be seen across the switch for a brief instant after the switch began to open. On a 24-volt circuit, the switch may have a positive 24 volts on the side toward the bus, and a negative 240 volts on the solenoid side. The arcing that occurs for an instant across the opening switch will conduct much of the negative 240-volt potential into the supply circuit.

13

Surge Suppression Devices and Designs

Surge suppression components and assemblies are relatively recent additions to the science of protecting against lightning's destructive effects. As pointed out in the preceding chapter, surge suppression technology has followed in the wake of ever increasing sensitivity of electronic devices to small and fleeting currents induced by lightning flashes, atmospheric electrification, and even by electrical switching.

Most manufacturers of equipment that can be affected by such surges provide limited surge suppression in the design of their equipment. But this suppression is often inadequate where exposure to transient currents is high.

According to Frank C. Breeze, associate principal and electronic systems consultant with the electrical engineering firm of Tilden, Lobnitz and Cooper, Inc. (Orlando, Florida), a surge suppression device must meet the following criteria to be of value to the user:

1. Under normal operating conditions, it must not interfere with the circuit it protects.
2. Its clamping voltage must not be greater than the surge withstand rating of the protected equipment.

3. Clamping speed must be fast enough to prevent damage to the protected equipment.
4. The device must be capable of withstanding surges without damage.

Frank Breeze has taught surge suppression and related subjects in seminars staged by the author at various locations in the United States as well as, on one occasion, in South Africa. He co-authored an article in the seventh edition of the National Association of Broadcasters' *NAB Handbook,* which encapsulates the essence of surge suppression and cites the advantages and disadvantages of various surge suppression components. The following is his advice.

The Primary Suppressor

The first line of defense against a powerline lightning or switching surge is the primary suppressor installed by the power company. These devices are installed between each phase conductor and the grounded neutral. They are commonly located at step-down transformers and at poles serving underground primary distribution (see Fig. 13.1). These devices are intended to clamp surges to a value below the insulation breakdown rating of power company transformers and cables.

The Secondary Suppressor

The second line of defense against surges is a suppressor installed at the building electrical service location (see Fig. 13.2). Most owners are familiar with the com-

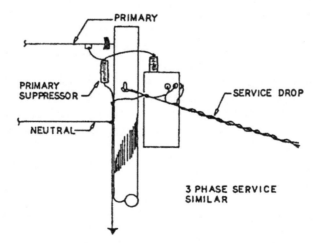

FIGURE 13.1 The local power utility provides each property owner with a primary surge suppressor, commonly located at a step-down transformer.

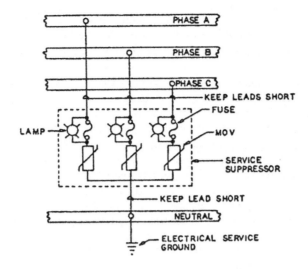

FIGURE 13.2 Next in line in virtually all installations is the secondary suppressor, which arrests overvoltages at the electrical service station.

mon cylindrical lightning arresters installed on commercial and residential electrical services. These devices consist of a low-voltage spark gap with a series thyrite element connected between each phase conductor and the grounded neutral. The thyrite element provides follow current limiting, allowing the arc to extinguish following a surge. This type of suppressor will clamp a typical surge at a level of 1,000 to 2,000 V, dependent on surge current and waveform. Clamping at this level is sufficient to prevent insulation breakdown on internal wiring, motors, and more rugged electrical devices.

A more effective device utilizes a large metal-oxide varistor (with protective fuse) connected between each phase conductor and the grounded neutral. Inductance of the connecting leads plays a major role in the effectiveness of the suppressor. The voltage drop through these leads is additive with the rated clamping voltage of the device. Suppressor leads should be as short as possible.

It is interesting to note that many codes do not require service surge suppressors on underground electrical services even though the service is fed from a pole a few feet away. It is advisable to provide service suppression in all applications.

Equipment Suppressors

It is quite possible to experience voltage surges of 2,000 volts within a building. These surges may be in the form of residual energy from an external event or generated within the building itself.

Equipment surge suppressors range in both complexity and in cost. These factors are usually in direct proportion to the clamping effectiveness of the device. Most of these devices depend on installation of a suppressor at the electrical service to prevent their exposure to large transients. One of the simplest and most common forms of equipment suppressors is a varistor and thermal fuse connected across the line and neutral conductors. Packaging is usually in the form of a series plug and socket arrangement or as a part of a plug strip. For a typical 2,000 V, 15 A surge, these devices will typically limit equipment exposure to 500 V.

For more sensitive equipment, hybrid suppressors are available that provide a lower clamping voltage and very fast response time. These devices use a high-energy metal-oxide varistor first stage, with the varistor and fuse connected in series between line and neutral. The line voltage then passes through a large core inductor to a silicon avalanche second stage, consisting of series connected bipolar suppression diodes between line and neutral. This configuration permits the varistor to absorb the majority of the surge while the diodes provide fast clamping. The inductor provides sufficient voltage drop during the surge to prevent damage to the diodes.

Surge Suppressor Components

Surge suppressor assemblies are packaged by a variety of manufacturers to protect a variety of circuits. To understand the operation of these devices, one must be knowledgeable in the characteristics of their component parts. The following section describes each basic type of device with its inherent advantages and disadvantages.

Spark Gaps

The spark gap consists of two electrodes placed in free air or some arc-quenching material. One important characteristic of the spark gap involves "flow current" when used on power circuits. Once an arc is established by a surge, the normal circuit voltage may be high enough to sustain the arc. This characteristic is normally dealt with by using a series interrupting device (circuit breaker), magnetic blow-out, series resistant element, or deionizer.

The advantages of a spark gap are that it offers:

1. Simplicity and reliability
2. High energy handling capacity
3. Very low voltage drop across the gap during conduction (typically 10 to 20 V)
4. Bipolar operation
5. Reasonably fast response

6. Zero power consumption
7. Long life expectancy
8. Low capacitance

There are also some disadvantages:

1. Used alone, it will not extinguish follow current.
2. It is limited to use on circuits of relatively high voltage.
3. Firing voltage depends on atmospheric conditions and surge rise time.

Gas Tubes

Gas tubes exhibit many of the characteristics of the spark gap (see Fig. 13.3). The problems associated with atmospheric influence are eliminated by enclosing the gap in an atmosphere of neon, argon, krypton, or other gas that is easily ionized at low pressure.

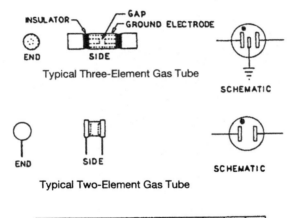

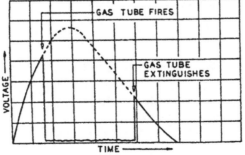

FIGURE 13.3 Gas tubes have gaps that are enclosed in an atmosphere of neon, argon, or other gas that ionizes easily under low pressure.

Advantages of gas tubes include:

1. Low cost
2. Small physical size
3. Good life expectancy
4. Fairly low capacitance
5. High energy capacity
6. Lower breakdown voltage than a spark gap
7. Very high current capacity and low clamping voltage

The disadvantages are:

1. Follow current limiting is required on power circuits.
2. Firing voltage depends on surge rise time.
3. It does not absorb appreciable surge energy.
4. It may be ionized by strong radio frequency (RF) fields.

Metal Oxide Varistors

Metal oxide varistors (MOVs) are composed of sintered zinc-oxide particles pressed into a wafer and equipped with connecting leads or terminals. These devices exhibit a nonlinear resistance characteristic and a more gradual clamping action than either spark gaps or gas tubes. As a surge voltage increases, these devices conduct more heavily and provide clamping action (see Fig. 13.4). Unlike spark gaps or gas tubes, these devices absorb energy during surge conditions.

The advantages of MOVs are:

1. They are available for low-voltage applications.
2. MOVs absorb energy.
3. No external follow current protection is required.
4. MOVs exhibit a fast response time.

The disadvantages are:

1. Clamping time depends on the surge wavefront.
2. External fusing is required for power applications (fails partially shorted).
3. MOVs have a limited surge life expectancy.
4. MOVs exhibit high capacitance.

Silicon Avalanche Devices

Silicon avalanche devices used for surge suppression are similar to Zener diodes except that they are designed to handle large surge currents without damage (see

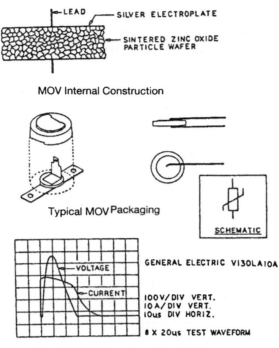

MOV Internal Construction

Typical MOV Packaging

Typical MOV Clamping Characteristic

FIGURE 13.4 Metal-oxide varistors (MOVs) are fast-response-time devices that require no follow current protection and, in effect, absorb energy.

Fig. 13.5). Junction construction for these devices is typically 10 times larger than an equivalent Zener device. The junction is sandwiched between silver electrodes to improve current distribution and aid in thermal dissipation. These devices exhibit an extremely fast clamping action with an absolute clamping level.

Silicon avalanche device advantages include:

1. High clamping speed (less than one nanosecond)
2. Hard clamping threshold
3. Availability for bipolar configuration
4. Small size

The disadvantages are:

1. They are subject to damage by large surges.
2. Lead length substantially affects clamping time.

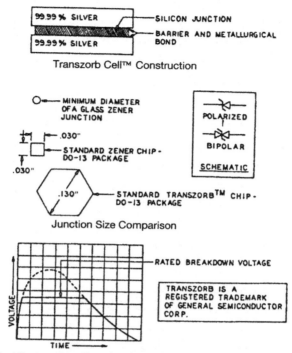

FIGURE 13.5   Silicon avalanche devices exhibit extremely fast response times and the ability to handle large surges without damage.

After reviewing characteristics of the basic surge suppression components, it is apparent that no single component is appropriate for all situations. It is, however, possible to use these devices in combinations to fit almost any surge suppression requirement.

14

Grounding the Thunderbolt

Wherever a lightning protection system is installed, all of the design and installation care invested in it can come to nothing if the contents and depth of the soil is not known and provisions are not made for any potential problems. Visualize, for example, that a common system layout is drawn for two identical buildings to be located only 50 miles apart in Massachusetts. One structure will be at Boston and the other at Worcester, to the west.

Being familiar with ground conditions in the Boston area, the designer lays out a minimum grounding system for each building, which for the low-rise structures consists of ground rods spaced at the maximum distance permitted under code—100-foot intervals. You specify 8-foot long, 0.5-inch diameter copper clad steel ground rods, driven 10 feet deep (see Fig. 14.1).

If lightning eventually struck the Boston structure, its powerful current would be accepted readily and dissipated quickly around the electrodes involved. Boston's soil has an extremely low resistance reading: only 0.5 ohms (Ω) at the first 10 feet of depth. If a lightning flash of the same magnitude struck the building at Worcester, severe damage and perhaps fire would be the likely result. The resistance to electrical current offered by Worcester's soil is extremely high—more than 3,000 Ω at 8 feet of depth, dropping to 30 Ω at a depth of 40 feet.

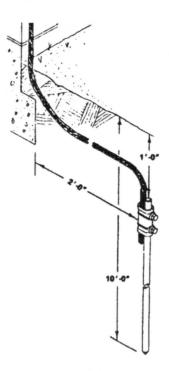

FIGURE 14.1 Minimum grounding in moist clay or loam consists of 8-foot ground rods, 10 feet deep, at 100-foot intervals.

THE TWO APPROACHES TO GROUNDING

America's consensus lightning protection standard, *NFPA 78,* does not define specific grounding requirements according to soil resistance readings. Instead, it classifies soils as (1) deep moist clay, (2) sandy or gravelly soil, (3) shallow top-soil, and (4) soil less than 1 foot deep. Minimum grounding requirements are established for each soil type. Figure 14.2, for example, shows a grounding configuration for gravelly soil. Australian, British, and South African codes, on the other hand, specify a ground resistance of less than 10 Ω for each down conductor, multiplied by the number of earth terminations, and require that the entire system have a combined earth resistance of no more than 10 Ω.

For industry-wide purposes, the American approach is the most practical, for two reasons. First, soil conditions vary widely in the United States, as illustrated in the Boston vs. Worcester example. Second, the degree of expertise possessed by designers and installers varies widely. From an overall standpoint, then, the specific and easily understood criteria offered by *NFPA 78* will assure good grounding in typical instances.

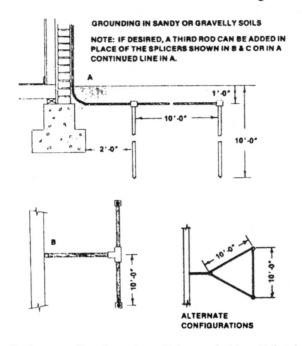

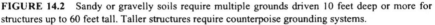

FIGURE 14.2 Sandy or gravelly soils require multiple grounds driven 10 feet deep or more for structures up to 60 feet tall. Taller structures require counterpoise grounding systems.

GROUNDING PROBLEMS AND SOLUTIONS

While the American approach is sound and appropriate for the lightning protection industry as a whole, there are situations where a more technical approach to grounding system design and installation can provide more positive protection, save installation time and expense, or both. Typical of common problem situations is a large site that contains an area of dumped rock, gravel, or other poorly conductive material. If a tall building requiring an encircling conductor or *counterpoise* is to be sited there, the ring conductor, combined with the division of down flowing current described in Chapter 10, will equalize ground resistance (see Fig. 14.3). However, if a large structure less than 60 feet tall is sited there, a counterpoise ground is not required. If lightning struck the system at the roof edge just above the poorly grounded rod, lightning current would have to travel to other down conductors and grounds, causing an increase in resistance.

SOIL VARIATION PROBLEMS

Another problem is misapplication of grounding standards due to lack of knowledge of the current-carrying capacities of soils that are similar in appearance. For

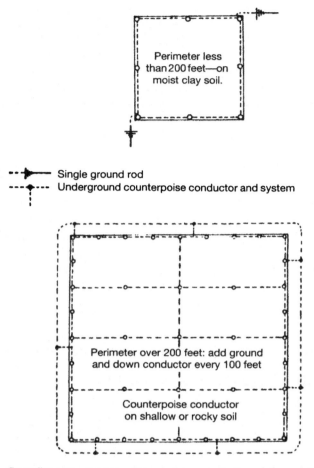

FIGURE 14.3 Grounding varies from two driven rods to an encircling conductor, which may be enhanced with driven rods or plates at the corners.

example, in Florida, an installer familiar with the low-resistance condition of salt-laden soil near the ocean or bay might be unaware that inland, where the soil appears to be similar, the ground water is quite deficient in conductive elements. In some locations, section rods driven to 100 feet or more of depth are required to achieve acceptable resistance readings.

SOIL RESISTANCE FACTORS

Five types of resistances need to be considered in achieving maximum efficiency in grounding lightning current. They are (1) type of soil, (2) depth, (3) moisture content, (4) temperature, and (5) the "transfer resistance" presented by the grounding media. It follows, then, that there are multiple opportunities to enhance grounding efficiency. Omitting temperature, which should be considered but cannot be controlled, there are four remaining areas where the lightning protection system designer and/or installing contractor can improve the efficiency of a grounding system.

Resistances of Different Soils

The U.S. National Bureau of Standards performed a series of tests on four categories of soils and has published the results appearing in Table 14.1. The Bureau's tests illustrate that rock or any rock derivative will degrade grounding performance according to the amount of rock present in the soil. Gumbo (the sticky-when-wet product of glaciation) has been found to offer the least amount of resistance. It is followed in ease of conductivity by clay which, like gumbo, tends to build up on the soles of shoes after a rain.

Table 14.2 (located at the end of this chapter) presents the results of nationwide soil resistivity tests and illustrates the broad disparity that exists within the United States and even within some individual states. Such disparity in readings at different locations underscores the importance of acquiring a knowledge of grounding principles and practices whether you are designing, installing, or inspecting lightning protection systems.

The following are examples of soil test locations where special measures may be necessary to achieve acceptable resistance readings.

Example 1. Earth resistivity tests at Mobile and Montgomery, Alabama, showed that high resistances there preclude reliance on conventional ground rods driven

TABLE 14.1 National Bureau of Standards Soil Test Results

Grounds Tested	Type of Soil	Resistance in Ohms		
		Average	Minimum	Maximum
24	Fills and ground containing varying amounts of refuse such as cinders, ashes and brine waste	14	3.5	41
205	Clay, shale, adobe, gumbo, loam, and slightly sandy loam with no stones or gravel	21	2.0	98
137	Clay, adobe, gumbo, and loam mixed with varying proportions of sand, gravel, and stones	93	6.0	800
72	Sand, stones, or gravel with little or no clay or stones	554	35	2,700

to 10-foot depths. The Mobile area readings, taken at the first 8 feet of depth, ranged from 700 Ω at one location to 3,000 Ω at the other. After a section rod was driven to a depth of 104 feet at the first location, the resistance reading dropped to 14 Ω. That is more than adequately low resistance for structural protection, but it indicates that it might be advisable to pay special attention to bonding for the sake of surge suppression. At the second test location, the reading after driving the test rod to 32 feet of depth dropped abruptly to a low of 6 Ω.

At site no. 2, the Montgomery test location, 800 Ω resistance was read at 8 feet of depth, and the reading dropped gradually to 120 Ω after the ground rod was driven to 120 ft., where it was stopped by impenetrable rock. A 120 Ω resistance reading is not acceptable even for structural protection, and special measures to further reduce resistivity should be considered.

Example 2. At Texarkana and Pine Bluff, Arkansas, resistance tests found acceptable conditions for structural protection at only 8 feet of depth, and very low readings were found at depths of 32 and 16 feet.

Deep Grounding Advantages

In lightning-prone Florida, where competition among installers is sometimes very stiff, diligent monitoring of grounding system installations may be in order. As indicated earlier, the presence of ground water at relatively high levels does not necessarily mean that earth resistivity will be low.

There are many other areas in the United States where deep grounding is either necessary or very desirable. For example, in some mountainous areas of Colorado, rocky conditions cause very high soil resistances. This is coupled with a high incidence of lightning flashes due to frequent orographic storms, in addition to frontal storms. These factors make positive lightning protection desirable, and deep drilling through rock to achieve acceptable resistance readings is appropriate.

The first few inches of soil are usually the driest and therefore the most resistant to the passage of lightning current. Obviously, that is not the case immediately following a rain, but at any other time, the first few feet of depth present unacceptably high resistance readings, as illustrated in Fig. 14.4. Note in the figure that earth resistivity in this example is reduced to the 50 Ω threshold of acceptability for structural protection at a depth of 4 feet. After that threshold is reached, the drop in resistivity slows down until it declines only 10 Ω between 7 and 11 feet.

The Advantages of Moisture

Moisture is essential to low earth resistivity as illustrated in Fig. 14.5. Note that in red clay, which is among the most conductive of soils, resistance is at 400,000 Ω when the moisture content is below 10 percent, and it drops precipitously to

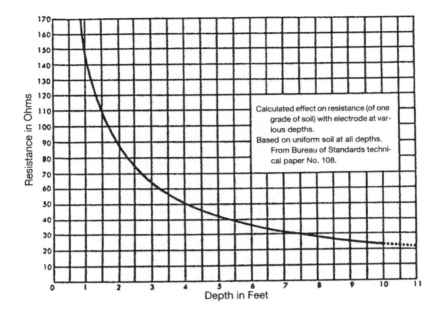

Calculated effect on resistance (of one grade of soil) with electrode at various depths.
Based on uniform soil at all depths.
From Bureau of Standards technical paper No. 108.

FIGURE 14.4   Soil resistivity varies with depth, sometimes widely. In dry weather, resistance to the acceptance of electric current is extremely high near the surface and still unacceptably high at 3 or 4 feet of depth. Earth resistance begins to flatten out at 3 to 5 feet in normal soil conditions.

20,000 Ω at 15 percent, then drops gradually toward single-digit readings as the moisture content increases.

Every locality has its wet years and dry years. In an area of average rainfall, variation ranges from 10 percent to 35 percent from year to year. Note in Fig. 14.6 that seasonal rainfall variations can make a difference of as much as 10 Ω from one time of the year to another.

How Temperature Affects Grounding

It is fortunate that thunderstorm activity peaks during the normally hot periods of the year when thunderstorms are usually at their most active stages. Freezing temperatures adversely affect earth resistivity levels, as shown in Fig. 14.7. For example, the graph shows a resistance to the passage of electric current of more than 30,000 ohms per cubic centimeter when the temperature hovers around 5° F.

The Effect of "Transfer Resistance"

When lightning current reaches a ground rod and flows along the rod's surface, it encounters a "transfer" resistance between the rod and the surrounding earth. This

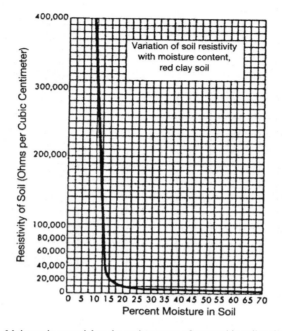

FIGURE 14.5 Moisture is essential to the maintenance of acceptable soil resistance levels. Note that a variation of only 5 percent in moisture content drops earth resistivity from 250,000 Ω at 10 percent moisture to about 20,000 Ω at 15 percent.

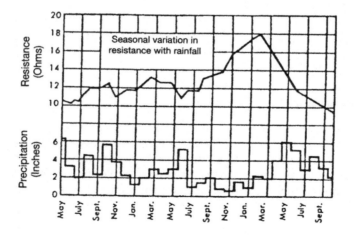

FIGURE 14.6 This chart, covering three years of resistance monitoring, shows that the normally wet seasons of spring and fall are the most favorable periods for lightning protection system performance.

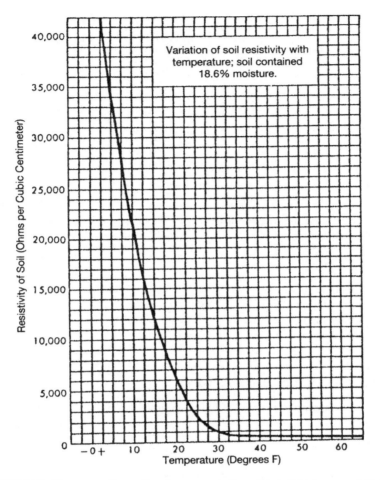

FIGURE 14.7 The warmer the weather, the more receptive soil is to lightning current. Note that when the temperature rises above freezing, the effects of temperature differences diminish drastically.

resistance continues and can be visualized as a series of shells increasing in size with greater distance from the electrode. A decrease in resistance with each succeeding shell continues until the expanse of earth is sufficient to accept the total amount of current (see Fig. 14.8).

Because the current transfer resistance decreases with distance, it follows that a small diameter rod will present a greater total resistance to the transfer of current from the rod to the immediately surrounding earth than will a larger diameter rod. However, experiments by Underwriters Laboratories, Inc., and the U.S. National Bureau of Standards showed that using a 1-inch diameter ground rod decreases the transfer resistance only about 10 percent from that presented by a 0.5-inch diam-

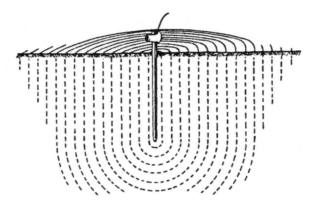

FIGURE 14.8 The earth's resistance to lightning current is most intense immediately surrounding the grounding electrode. It diminishes gradually in a circular pattern around the electrode. The diameter of the rod plays a minor role in transfer resistance. The rods should be selected on the basis of resistance to bending.

eter rod. Because the difference is so small, the Copperweld Company, a leading manufacturer of ground rods and related products, advises that ground rods should be selected on the basis of resistance to bending and adherence to code requirements rather than to the effects of transfer resistance.

TYPES OF GROUNDING MEDIA

There are essentially eight types of lightning protection grounding media:

1. Driven rods
2. Extended conductors
3. Counterpoise (encircling) conductors
4. Ground plates
5. Ground reservoirs
6. Ufer grounds
7. Mesh grounding
8. "Convenience" grounds

Driven rods are available in copper, copper-clad steel, and stainless steel. Lengths range from a minimum of 8 feet to lengths as long as 20 feet. Section rods are used for deep grounding. The most common diameters are 0.5, 0.625, and 0.75 inches.

Copper-clad steel is by far the most commonly used material. Where chemicals, manure, or other corrosive material exists in the soil, stainless steel rods are used because of their greater resistance to corrosion.

Extended cable conductors of Class I dimensions and at least 12 feet long are laid in trenches, projecting outward from the structure. They are used where high bedrock precludes the use of driven rods. Their length may be limited to 12 feet in conductive soils such as clay, gumbo, or loam that is high in clay content. If the soil is other than those named, the trenched conductors must be at least 24 feet long and may need to be enriched.

Counterpoise conductors encircling the building or other structure are required for all structures where the soil is less than 1 foot deep. Such encircling conductors are also required by *Code NFPA 78* for all structures that are more than 60 feet in height.

Because a minimum of cutting is required, the least expensive twisted copper cable is commonly used for counterpoise grounding as well as trenched, extended conductors. Stainless steel conductors should be considered in soils that appear to be corrosive to metals.

Copper plates are permissible under restrictive conditions as alternate grounding devices. A copper grounding plate must be at least 0.032 inches thick and 2 feet square. Such plates should be viewed as last-resort alternatives. Their more appropriate role is as an additional grounding device at the corners of structures standing on soils less than 1 foot in depth.

A *ground reservoir* is a cavity in the earth, filled with scrap metal, charcoal, vermiculite, or other conductive material. It is usually located in a low spot where moisture collects. The ground reservoir offers a solution to grounding where no other viable alternative exists. If there is an opportunity to create a ponded reservoir in the vicinity, it can serve as a ground for a structure sited on bare rock, as an alternative to drilling through such rock to reach conductive earth.

A *Ufer ground* consists of copper conductors encased in a building's footings, where they are located near the bottom of the footings. Named for the man who proposed such grounding, Ufer grounds work on the principle that concrete under compression will absorb and maintain a moisture content and serve to transfer lightning's electric current to earth along the length of that conductor.

Mesh grounding is sometimes used where no sparks can be tolerated, whether caused by lightning or by friction. It consists of a wire network in contact with earth, either directly or by virtue of close proximity through location near the bottom of the building's concrete slab.

Convenience grounds are buried metal pipes and other underground metal bodies that can be utilized as part of a lightning protection system. A metal waterline leading to a metal well casing makes an excellent ground terminal, for example. So does a metal waterline or waste line provided by a municipality. Such lines should be viewed as individual ground terminations, not as complete grounding systems.

DEEP GROUNDING TEST RESULTS IN THE
UNITED STATES

The Copperweld Company has published a tabulation of results obtained in deep grounding tests at 219 locations in the 36 most lightning-prone states in the United States (see Table 14.2). Eight states—Alabama, Connecticut, Florida, Massachusetts, Mississippi, New York, Rhode Island, and Wisconsin—had locations where resistance readings topped 3,000 Ω. Wellsburg, West Virginia, reported only 1 Ω of earth resistance, while Johnstown, Pennsylvania, had 2 Ω, and Findlay, Ohio, had 2.5 Ω.

TABLE 14.2 Deep Grounding Test Results

Location	$A*$ $\Omega\dagger$	$B\ddagger$ $\Omega\dagger$	B Feet**	Location	$A*$ $\Omega\dagger$	$B\ddagger$ $\Omega\dagger$	B Feet**
Alabama				**Iowa**			
Atmore	1,000	47	96	Boone	6	3.5	25
Birmingham	80	7	31	Brooklyn			
Birmingham	130	19	13	(Williamsburg)	10	4.5	32
Florala	950	6	73	Council Bluffs	15	3.5	16
Foley	2,900	33	120	Creston	8	3.5	24
Mobile	700	14	104	Davenport	7	6	24
Mobile	>3,000	6	32	Grinnel	12	3.5	24
Montgomery	800	120	69	Hampton	45	10	24
Arkansas				Mason City	76	25	32
Texarkana	15	2	32	Muscatine	2,000	8	96
El Dorado	50	12	89	Newton	13	3.5	24
Pine Bluff	12	2	16	Sioux City	30	7	64
				Vinton	16	7.5	17
Connecticut				Waterloo	165	4	64
Milford	265	40	15	Wilton Junction	11	4.5	48
Windsor	>3,000	17	104	**Kansas**			
Florida				Great Bend	42.5	4	14
Daytona Beach	240	50	72	Topeka	10.5	3	16
De Funiak Springs	1,000	9	65	Wichita	7.5	2	40
Fort Myers	32	3.5	32	**Kentucky**			
Interlachen	>3,000	10.5	113	Ashland	30	1	64
Jacksonville	225	19	56				
Live Oak	450	4	33	**Louisiana**			
Melrose	600	8.5	80	Alexandria	82	14	32
Panama City	1,800	19	64	Baton Rouge	11	1	24
Pensacola	20	5	40	Homer	58	6	56
Qunicy	2,700	6	88	Lake Charles	4	1.5	16
St. Marks	5	2	13	Natchitoches	48	5	16
Tampa	100	6	22	New Orleans	200	5.5	64
Waldo	150	3.5	41	Shreveport	2	1.5	19
West Palm Beach	150	17	32	West Monroe	135	26	96
Georgia				**Maine**			
Atlanta	400	85	40	Augusta	2,500	1,300	28
Elberton	21	15	24	Bangor	32	18	30
Illinois				Portland	2,100	550	22
Alton	210	6	72	**Maryland**			
Decatur	11.5	7	30	Hagerstown	15	11	14
Dixon	150	8	70	**Massachusetts**			
Maywood	8.5	4	30	Boston	0.5	0.5	30
Peoria	32	4	37	Lowell	910	35	60
Rockford	26.5	4.5	80	Manomet	2,000	18.5	70
Springfield	4	1.5	30	Springfield	>3,000	20.5	48
Indiana				Worcester	>3,000	39	40
Evansville	5.5	4.5	30	**Michigan**			
Fort Wayne	22	12	30	Detroit	13	2	20
Hammond	8	4	30	Jackson	250	6	30
Indianapolis	21	7	30	**Minnesota**			
Marion	6.5	4.5	20	Alexandria	200	6	41
South Bend	26	6	30				

TABLE 14.2 (continued)

Location	A* Ω†	B‡ Ω†	B Feet**	Location	A* Ω†	B‡ Ω†	B Feet**
Minnesota (cont.)				**New York**			
Anoka	1,750	18	64	Binghamton	47	26	48
Aurora	70	44	**18**	Elmira	13.5	7	48
Bemidji	2,800	7	48	Holbrook	>3,000	100	127
Battle Lake	225	11	64	**North Carolina**			
Brainerd	500	5	33	Charlotte	500	51	**18**
Duluth	45	9	40	Raleigh	290	19	**61**
Elk River	460	10.5	64	**North Dakota**			
Minneapolis	29.5	1.5	40	Bismarck	350	4	33
Owatonna	195	13	30	Devils Lake	16	4	16
Slayton	15	5	21	Dickinson	270	12	23
Virginia	2,800	36	128	Grand Forks	16	3	16
Willmar	500	19	71	Jamestown	35	8	23
Mississippi				Valley City	35	6	16
Bay St. Louis	100	5	64	Velva	475	14	33
Bay St. Louis	700	2	40	Williston	280	15	23
Columbia	150	2.5	24	**Ohio**			
Greenwood	18	2.5	24	Canton	24	19	17
Iuka	750	7	64	Cincinnati	6	5	30
Jackson	550	7	96	Cleveland	18	8	16
Laurel	170	5	24	Dayton	70	21	35
Lucedale	>3,000	59	96	Findlay	2.5	2.5	16
McComb	400	1	56	Massillon	18	18	17
McComb	40	2.5	32	New London	11	7	30
Meridian	100	5	40	Toledo	7	3	30
Meridian	200	50	106	Warren	6	8	20
Natchez	26.5	8	40	Youngstown	35	6	40
Picayune	100	2.5	88	**Oklahoma**			
Pontotoc	1,300	11	32	Blackwell	10	4	21
Taylorsville	165	110	40	Oklahoma City	23	13	**18**
Tishomingo	265	5.5	48	Tulsa	110	11	22
Utica	40	7	17.5	**Pennsylvania**			
Wiggins	1,750	7	96	Bedford	21	18	
Yazoo City	140	6	56	Brookville	155	45	
Missouri				Gibsonia	74	11	
Jefferson City	7.5	5.5	12	Greensburg	58	11	
Joplin	3	2.5	24	Indiana	30	29	
Kansas City	14.5	2.5	48	Johnstown	2	1.75	
Warrensburg	11.5	5.5	**16**	Kittanning	13	5	
Montana				New Castle	350	75	
Glendive	205	10	39	North East	195	14	
Glendive	40	3	23	Parker's Landing	95	78	
Nebraska				Philadelphia	320	240	
Lincon	6	5.5	16	Reading	92	33	
Omaha	23	6.5	32	Somerset	82	21	
Valentine	350	15	30	**Rhode Island**			
New Hampshire				Providence	>3,000	275	32
Manchester	38	5	40	**South Carolina**			
Manchester	6	3.25	20	Columbia	1,400	160	**22**

TABLE 14.2 (continued)

Location	A* Ω†	B‡ Ω†	B Feet**	Location	A* Ω†	B‡ Ω†	B Feet**
S. Carolina (cont.)				**Texas (cont.)**			
Columbia	1,150	11	96	Denton	3.5	19	24
Ellenton	225	1.5	88	Douglassville	200	27.5	**45**
South Dakota				Greenville	2.5	.75	23
Hudson	700	42	**41**	Jacksonville	36	9	**32**
Huron	65	4	19	Plainview	100	3.5	40
Ipswich	3	2	30	San Antonio	10	3	22
Mitchell	22	4	19	Tyler	235	4	24
Philip	4	4	26	Waco	17.5	1.5	24
Rosen Church	130	9	31	**Virginia**			
Salem	4	3	20	Chase City	170	55	35
Vermillion	150	6	40	Hairfield	1,000	310	**20**
Watertown	100	6	31	Paytes	325	85	48
Yankton	20	5	32	Petersburg	400	4	62
Tennessee				Roanoke	325	110	**24**
Bristol	37	11	54	Salem	18	12	**23**
Chattanooga	525	60	**22**	Bowling Green	700	175	46
Chattanoga	175	18	**51**	**West Virgnia**			
Greenback	340	60	44	Bluefield	11	10	**16**
Jackson	110	10	48	Charleston	19	7	37
Knoxville	800	35	**100**	Huntington	39	18	71
Memphis	42	11	56	New Haven	15.5	6	80
Texas				Wellsburg	1	1	16
Abilene	62	5	22	**Wisconsin**			
Amarillo	19	4.5	54	Ashland	125	24	36
Athens	70	32	27	Barron	2,800	24	64
Beaumont	4.5	1.5	24	Darlington	11	10	**9**
Bells	5	5	**23**	Eau Claire	>3,000	21	98
Corpus Christi	400	3	24	La Crosse	1,900	10	112
Corsicana	3	2	12	Oakdale	52	54	**19**
Dallas	3	6	24	Milwaukee	7	4	20
Denison	14	8	**16**				

* First section—usually 8 feet
† Resistance in ohms
‡ Deep ground—depth shown in third column
** Bold = underlying rock prevented deeper driving

15

Lightning Warning and Locator Systems

During the 1991 U.S. Open golf tournament at Hazelton National Golf Club, in Chasca, Minnesota, lightning struck and killed a spectator and injured five others. After the tragedy, newspapers printed their reports under such headlines as "golf must take steps to warn fans of storms" and "lightning detector no guarantee of golf tournament safety."

The first headline was appropriate, the last questionable. Among the professional golfers playing in the tournament was well-known pro Lee Trevino, who had once survived a painful scrape with lightning. Trevino, Jerry Hurd, and Bobby Nichols suffered severe jolts of ground voltage as they leaned on their clubs some distance from a lightning strike to ground during the 1975 Western Open tournament at Chicago.

At Chasca, Trevino finished play on the fifteenth hole, chipping from the fringe and sinking a putt before seeking shelter. Asked why he ignored the warning and continued to play in defiance of the imminent danger communicated by the wailing sirens, Trevino commented, "We are just as dumb as they are (the spectators); we think it won't strike us."

TRADITIONAL WARNING UNITS

Field mills were the first units to be used by golf courses, mining operators, and blasting crews. They are downward-looking units that measure atmospheric field strength. Three separate mills located miles apart can locate a thundercell and send alarms to users. The alarms may be visual, audible or both.

THE NEW LIGHTNING LOCATOR
METHODOLOGY

Lightning Location and Protection, Inc., (LLPI) is a company located in Tucson, Arizona, that claims to " . . . provide a unique source of accurate and timely information on the presence or absence, location, intensity, and movement of lightning storms over areas of virtually any size." (See Fig. 15.1) After the 1991 U.S. Open golf tournament at Hazelton National Golf Club, near Minneapolis, Minnesota, LLPI analyzed the sequence of events that led to the tragedy. For the analysis they used a lightning stroke detection and location information obtained from the Lightning Position and Tracking System (LPATS) arm of the company.

LPATS tracked three centers of atmospheric action in the area—one to the north, one to the northeast, and one to the south. The northerly and northeasterly thundercells had reached maturity minutes before the tragedy. There was no evidence to indicate that the fatal strike was a "bolt from the blue," where a lightning flash travels horizontally for many miles before stabbing to ground. In LLPI's scenario no. 1, a weak convective thundercell developed over the course in the area of the eleventh tee. This fourth cell was seen as the source of the fatal strike.

In scenario no. 2, the center of atmospheric action to the south of the golf course continued to expand, and the fatal bolt came from that cell. The conclusion was that scenario no. 2 was " . . . reasonable since there was a mirror image of the activity in the area north of the course, and the displaced flash sequence falls within a reasonable distance of the course and the axis of cell trajectory, especially if validity is given to scenario no. 1 and the fact that an increase in the electric charge in the atmosphere was already appearing over the course."

Lightning's Radio Impulses

The basic principle on which the new lightning locating system operates is that each upward-traveling return stroke produces a very strong radio frequency impulse that travels outward at nearly the speed of light and can be detected for hundreds of miles. The peak radiated power of an average return stroke is much stronger than any man-made radio transmitter can produce.

The radio signal produced has a very characteristic signature that is different from the radio signal of intracloud or cloud-to-cloud lightning. The signal is also

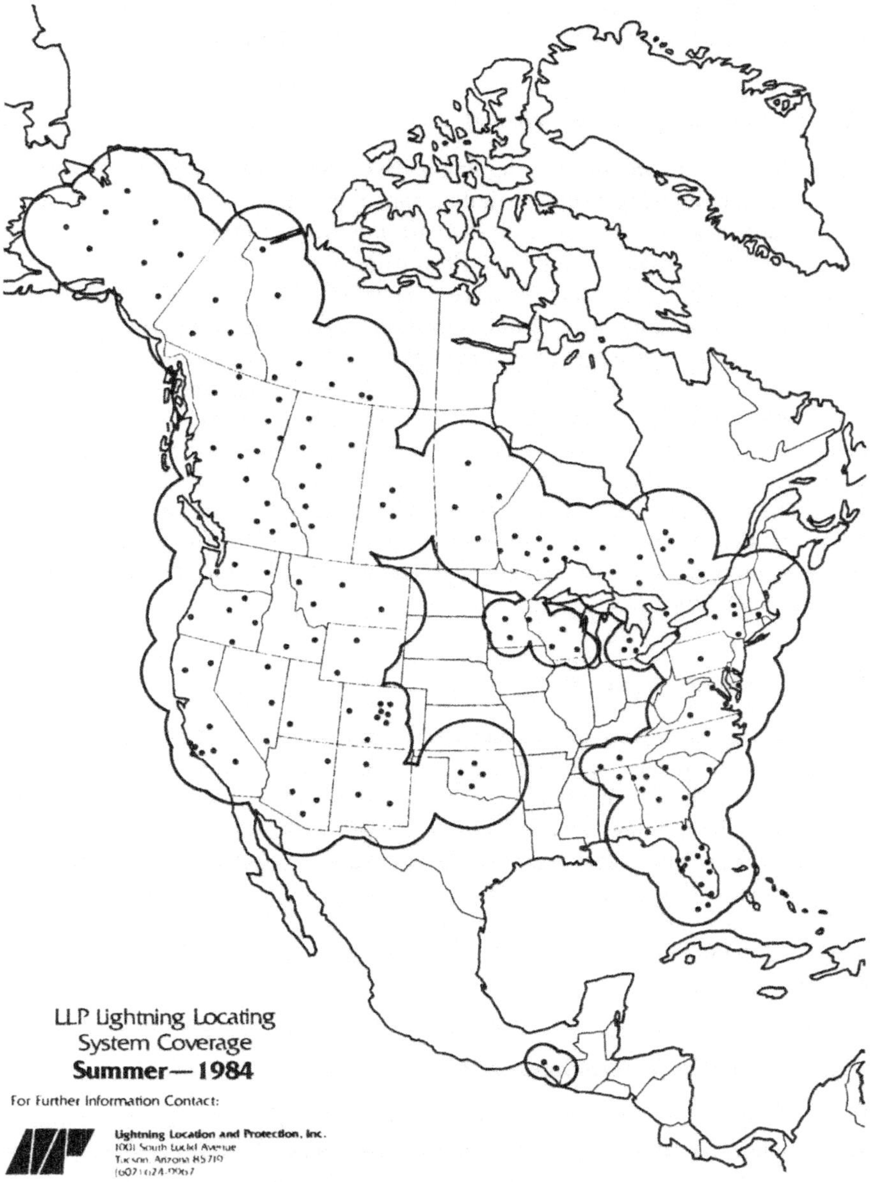

LLP Lightning Locating
System Coverage
Summer—1984

For Further Information Contact:

Lightning Location and Protection, Inc.
1001 South Lucid Avenue
Tucson Arizona 85719
(602) 624-9967

FIGURE 15.1   This is the area of coverage indicated for North America, as Lightning Locating and Protection, Inc., provided it in 1984. The company says it extends coverage only when it can do so while maintaining its quality level.   Reproduced by permission of LLPI.

different from that produced by man-made background noise. It is this singularly different signal that is picked up by an LLPI lightning locating system.

How Lightning's Signal Is Detected

All radio impulses produced by lightning flashes in an area of operation are received at each lightning direction-finding station in that area at the same moment. Special antennas and electronic measurement devices determine the direction to the ground strike point of a lightning flash with great accuracy.

Direction-finder electronics analyze the characteristics of radio impulses and reject all impulses that are not from a lightning strike to earth. Since a number of return strokes from different flashes may occur virtually at the same moment, the direction-finder electronics identify each stroke on the basis of time and angle consistency. In a few thousandths of a second, each direction finder transmits a brief message to the position analyzer. The message contains the exact location, polarity, signal amplitude, and number of return strokes in each cloud-to-ground flash. The message is normally transmitted by telephone circuits leased from the local telephone company. The position analyzer calculates the exact coordinates of the lightning locations by triangulation and displays the time, angle, and range to the ground strike.

Displaying the Lightning Data

After each lightning flash has been processed in its split second of time, the position analyzer transmits the time, latitude, longitude, polarity, and intensity characteristics of the flash to any number of remote display processors (also by phone line). When the data are received, a high-resolution color graphic display shows both the current and recent lightning strike locations superimposed on user-defined background map information (see Figs. 15.2 and 15.3). The map region and scale of the display may be easily changed to view different areas, and a limited amount of lightning data may be stored locally for recall of the last several days of lightning data

Golf Course Safety Recommendations

Because golf courses gather numbers of people who may be exposed to lightning dangers, there has been momentum by course management toward assuming some responsibility, plus corresponding movement in the legal community toward fixing responsibility on them. For normal course occupancy, shelter buildings equipped with lightning protection systems should be located within reach of golfers on the course. The clubhouse, also protected, should house either a warning system hooked to a field mill or a lightning location and warning service. During

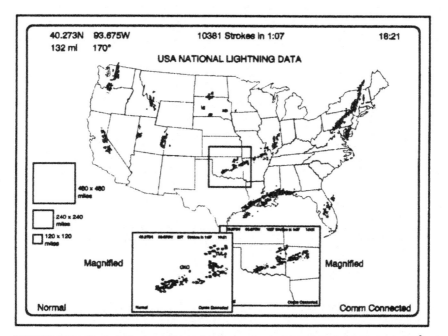

FIGURE 15.2 Typical Display of Lightning Storms by the LPATS National Network (LN^2). A high-resolution color graphic display shows the location and area of a storm. The map region and scale of the display are easily and quickly changed as required. Courtesy of Atmospheric Research Systems, Inc.

a golf tournament, the size of the gallery usually greatly exceeds the number of golfers and their support personnel. For that reason, Lightning Location and Protection, Inc., experts recommend that there be two distinct programs.

The first step is to establish conditions of readiness (CORs), assigning an initial COR that is applicable to both golfers and spectators. Time frames need to be established that identify (1) a period of time when people need not concern themselves with the threat, (2) a time frame when it can be anticipated that a problem will evolve, and (3) the time span required for people to find safe shelter.

Either of two statements would be issued based on weather conditions:

1. *Lightning Condition II.* "Based on local weather forecasts, thunderstorm activity is forecast for the area during the scheduled period of play. These thunderstorms may have an impact on play if they appear as forecast. However, no activity is forecast for at least two hours." The message does not impact conduct of the tournament, but it does increase the awareness of all parties that they may have to deal with thunderstorms during the period of play.

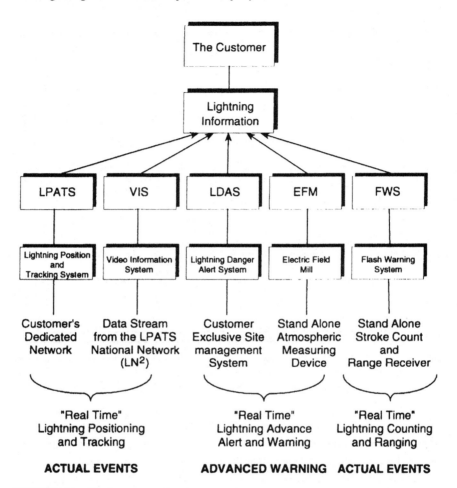

FIGURE 15.3 The five systems, from location of the lightning storm and its movement to counting of the strokes, are indicated in this figure. Courtesy of Atmospheric Research Systems, Inc.

2. *Lightning Condition I.* "Based on local forecasts and available data, thunderstorm activity is present or expected to develop within the local area and will have an impact on play within the next two hours." This second COR statement alerts people that there will be thunderstorm activity, based on available data. There will be some impact on play, and the statement prompts people to think about where they will go if the storms come.

Lightning warnings are tailored to the specific needs of the golfers and spectators, depending on course distances, proximity of automobile parking and difficulty of vacating spectator areas.

Lightning Alarms

Most golf courses are equipped with sirens or air horns. While these are effective alarms, it is important that the alarm used for spectators be distinctly different from the one used for golfers and not serve as a source of distraction to the golfers.

If a siren is used, the length of its run should not exceed 10 seconds. There should be a 30-second break between sirens, and the 10-second siren alarm should be repeated at least five times. If an air horn is used, two pulses, followed by a 20-second pause, would represent one cycle. The cycle should be repeated at least five times.

In a lightning warning stage, lightning is imminent or expected within the next 15 minutes. All play is suspended. All people are to leave the course immediately and seek safe shelter, and it is anticipated that the current threat will exist for at least one hour. People should not return until the all-clear signal is sounded.

By issuing this warning, action is initiated to protect the golfers and, in essence, people are being told (not asked) to leave the course. As with the spectator warning, audible and visual alarms should be used. Since the red light was already activated during the spectator warning, no distinct visual alarm is needed. The alarm should be prolonged, thus leaving no doubt that the threat is present.

Once the threat is over, a decision must be made as to whether other weather activity may still pose a threat or if there will be no further weather problems for the remainder of the day. If the latter is the case, an "all clear" status would return everyone to the course with the understanding that no thunderstorms will impact on play. Such an action would be typical after a front or squall line has passed through the area.

On the other hand, when dealing with air mass thunderstorms that occur early in the day or are forming in the area in a cyclic pattern, it is possible that conditions will warrant returning to a condition of readiness or spectator warning level of readiness due to the continued possible threat.

16

Standards, Inspections, and the Engineer's Role

Engineers, particularly those engaged in the electrical and electronic disciplines, are becoming more involved in the lightning protection arena, and that is as it should be. While the large majority of lightning protection manufacturers and installing contractors are quality conscious, they and their customers can benefit from a knowledgeable engineer's hands-on participation in the specifying and monitoring processes.

Traditionally, lightning protection has been a competitive field where installation speed and frugal use of materials are vital to realization of a proper profit. Consequently, most protection systems meet minimum code requirements, but do not go beyond that to provide the margin that might be desirable for a building housing a multiplicity of sensitive equipment.

LIGHTNING PROTECTION STANDARDS

Four organizations are involved in lightning protection system quality control in the United States. They are:

1. *The National Fire Protection Association (NFPA).* NFPA provides the organizational structure and staff facilities for the Lightning Protection Code Committee.

2. *The Lightning Protection Institute (LPI)*. LPI was created by manufacturers of lightning protection system components and since has added contractor and engineer members. It supports the use of lightning protection systems through a public relations program.

3. *Underwriters Laboratories, Inc. (UL)*. This organization monitors quality in both the manufacture of system components and installation.

4. *The American National Standards Institute (ANSI)*. ANSI was created to serve as an umbrella organization, in effect an association for not-for-profit associations throughout the United States.

Code NFPA 78

The National Fire Protection Association first adopted the *Specifications for Protection of Buildings Against Lightning* in 1904. Revised standards were adopted in 1905, 1906, 1925, 1932, and 1937. In 1945, the NFPA Committee and the parallel ASA Committee on Protection Against Lightning were reorganized and combined under the sponsorship of the NFPA, the National Bureau of Standards, and the American Institute of Electrical Engineers (now the IEEE). In 1946, the NFPA acted to adopt part III and in 1947 published a revised edition incorporating this part. Further revisions recommended by the Committee were adopted by the NFPA in 1949, 1950, 1951, 1952, 1957, 1959, 1963, 1965, 1968, 1977, 1980, 1983, 1986, and 1989.

The *Lightning Protection Code, NFPA 78,* covers protection for ordinary buildings, miscellaneous structures and special occupancies, heavy-duty stacks, and structures containing flammable liquids and gases. It does not cover lightning protection requirements for explosives manufacturing buildings and magazines or electricity generating, transmission, and distribution systems.

The code is written by the Technical Committee on Lightning Protection, which is composed of roughly equal numbers of representatives from three areas: the lightning protection industry, the engineering and scientific fields, and users. The latter may be involved in military, industrial, or insurance fields, for example.

A committee chairman is appointed who, in turn, appoints a secretary who keeps minutes of meetings and assists in various other ways. Committee members name alternates who may attend meetings and may vote in the member's absence.

A *Technical Committee Documentation* is published annually. This publication indicates the actions taken by each Technical Committee during the prior annual period that results in a technical change to their standard. Such changes are considered at the NFPA Annual Meeting, where they are considered for acceptance, rejection, or a recommendation for alteration and resubmittal.

One notable action by the Lightning Protection Code Committee during the author's long tenure as its secretary involved a proposed requirement for lightning protection systems for explosives manufacturing and storage facilities. This drew several pages of critical comments from within the explosives manufacturing and user industries, whose members insisted that there already were adequate rules and policies in that area, and that standard lightning protection systems were not needed.

Published in the 1989 *Technical Committee Documentation* manual for consideration at the 1989 annual meeting, the comments were duly considered, and the outcome was a diplomatic action adopting the code committee's proposal but placing the section in the 1989 edition of *NFPA 78* as a non-mandatory proposal, Appendix L, "Protection of Structures Housing Explosive Materials."

The 1989 edition of *NFPA 78* has six chapters in its mandatory section. They are as follows:

Chapter 1   Introduction
Chapter 2   Terms and Definitions
Chapter 3   Protection for Ordinary Structures
Chapter 4   Protection for Miscellaneous Structures and Special Occupancies
Chapter 5   Protection for Heavy-Duty Stacks
Chapter 6   Protection for Structures Containing Flammable Vapors, Flammable Gases, or Liquids That Can Give Off Flammable Vapors

The appendices are:

Appendix A   Introduction
Appendix B   Inspection and Maintenance of Lightning Protection Systems
Appendix C   Guide for Personal Safety During Thunderstorms
Appendix D   Protection for Sailboats, Power Boats, Small Boats, and Ships
Appendix E   Protection for Livestock in Fields
Appendix F   Protection for Picnic Grounds, Playgrounds, Ball Parks, and Other Open Places
Appendix G   Protection for Trees
Appendix H   Protection for Parked Aircraft
Appendix I   Risk Assessment Guide
Appendix J   Ground Measurement Techniques
Appendix K   Explanation of Bonding Principles
Appendix L   Protection of Structures Housing Explosives
Appendix M   Principles of Lightning Protection
Appendix N   Referenced Publications

The National Fire Protection Association is located at Batterymarch Park, Quincy, MA 02269.

UL's Role in Lightning Protection

Underwriters Laboratories, Inc., began serving property owners in the lightning protection field in 1908 and now has "a large number of trained lightning protection field representatives located throughout the United States." Such representatives inspect sites ranging from cow barns to missile silos, and from golf course shelters to high-rise building systems. According to UL literature, the White House, Washington Monument, and the Sears Tower in Chicago (the tallest building in the world) have metal UL Master Label plates attached to them.

Services available to the lightning protection industry from Underwriters Laboratories entail the entire gamut of products and services. They are as listed below.

- Factory inspections on an unannounced basis of conductor cable, ground rods, connectors, clamps, plates, air terminals and their bases, and other parts are carried out from four main offices. They are at 333 Pfingsten Road, Northbrook, Illinois; 1655 Scott Boulevard, Santa Clara, California; 12 Laboratory Drive, Research Triangle Park, North Carolina; and 1285 Walt Whitman Road, Melville, New York.
- UL offers in-house inspection and testing of electric service and telephone surge arresters.
- UL provides testing and listing of transient voltage surge suppressors used for protection of fragile electronic equipment.
- UL offers a listing of installers of lightning protection systems.
- UL provides field inspections of installed lightning protection systems.

When an engineer or other authority specifies conformance to UL requirements under the Master Label program, the owner signs the owner's statement on the application form and receives a copy of the form for his records. After the installer submits the Master Label application to UL, the metal label is issued to him for placement by the owner or installer on the building—usually near a down conductor.

Buildings that are changed structurally or provided with additions can be reexamined under the UL Reconditioned Lightning Protection Program. To qualify, the entire system must comply with current UL standards.

The LPI and Its Programs

Several years ago, spurred by complaints about Underwriters Laboratories inspections, the Lightning Protection Institute acted to create its own inspection and certification program. The program was designed to provide a building owner with a procedure for complete and timely inspection of the system installation.

Two of the criticisms by lightning protection installers that prompted the Institute to create its own quality control program were:

1. UL inspectors usually do not arrive until the installation is complete. Because the system's down conductors are usually encased within the walls for appearance's sake, the UL inspector may have difficulty determining that the roof system is indeed connected to the grounding system.
2. Inspection of a UL-listed installer's work is done on a spot basis, and many installations are therefore not viewed by a knowledgeable third party.

Under the LPI program, a complete inspection form that requires viewing of the various components is used. For residential or farm installations, the house or farm owner may make the inspection, using the form and a copy of the Institute's liberally illustrated *Standard of Practice LPI 175*.

For commercial, industrial, and institutional structures, the inspector may be the electrical engineer involved in the job, a clerk of the works, the general contractor's foreman, or another responsible third party individual.

The Engineer's Role

As stated at the beginning of this chapter, engineers—particularly those involved in the electrical and electronic disciplines—have become increasingly involved in lightning protection in recent years. Evidence of this trend has been displayed by attendance at seminars on lightning protection system design, installation, and inspection. The majority of attendees at the author's School of Lightning Protection Technology seminars, for example, have been electrical engineers. It is common for larger engineering firms to have one or two people involved in the lightning protection arena.

However knowledge of lightning protection system design and inspection is gained, a professional engineer can benefit the supplier of lightning protection goods and services as well as the customer for those services by serving as a knowledgeable intermediary. As technology continues its march, the user's need for such professional services will also continue to expand.

17

Ineffective Devices

As a complex, threatening, and to many people mysterious celestial force, lightning has long been challenged by ineffective actions and instruments. Centuries ago, such challenges were made in ignorance and innocence.

As cited in Chapter 1, it was in 1752 when Benjamin Franklin flew his famous kite into an electrically charged atmosphere as lightning flashed and thunder crashed and rumbled in a blackened sky. In this era, the so-called "sanctified" tones of church bells were still climbing the airwaves to combat the supposedly evil noise of thunder. And here and there village cannons boomed as townsfolk sought to overcome thunder with louder racket.

Today church bells ring only to call worshippers together, and the few cannons that still grace village squares no longer boom challenges skyward on stormy days. But ineffective and therefore dangerous contrivances and methods are still being foisted on an unwary public, and these days not in ignorance but with practiced, deceptive intent.

THE FECKLESS TRIO

Several alternative devices are claimed by promoters to be effective substitutes for modern lightning protection systems that adhere to the requirements of the National Fire Protection Association's *Lightning Protection Code NFPA 78*. The three general categories are:

1. *Radioactive lightning rods.* These devices continuously produce upward-moving emission currents as the radioactive isotopes decay. The rate of decay varies according to the type of radioactive material employed. Decay time periods range from seconds to hundreds of years among nature's many isotopes.
2. *Lightning dissipation arrays.* These systems consist of hundreds of sharp metal points. It is claimed that lightning will not strike buildings equipped with such arrays.
3. *"Pulsing" air terminals.* Promoters of these devices claim that applying a pulsed high voltage to an air terminal's tip will give positive ground currents a boost upward toward descending negative leader strokes, thus reaching out far beyond the attractive range of conventional lightning rods.

Radioactive Lightning Rods

Radioactivity is a natural process by which atoms emit atomic particles and rays of high energy. Certain radioactive materials are used by physicians to destroy cancer cells. Other such materials are employed in research and in industry.

Radiation can be extremely harmful to skin and tissue, or it can be relatively benign, depending on its *half life*. The term signifies the amount of time it takes for the material to decay to half its original value.

Promoters of radioactive lightning rods claim that the radioactive emission gives a powerful extra boost to ground-based electric currents which are continually produced in a natural interchange between earth and the atmosphere. However, atmospheric scientists in America, Europe, and South Africa, as well as in South America, Australia, and the Orient, have dismissed such claims out of hand. To them, and to others who are knowledgeable about lightning, the claim that any substance or device that would be made available at acceptable price could match the natural attraction between positive earth and a negative thundercloud is ludicrous.

Radioactive lightning rods have been banned in several countries. One of America's leading atmospheric scientists, Dr. Rodney Bent, explains why in a statement written at the author's request:

> In order to examine the claims related to radioactive lightning rods, it is necessary to consider the physical process of a discharge to a conventional rod. When a lightning rod is in the area of a lightning leader, the electric field around its tip would be extremely high and the air in this region would be in glow discharge, which means millions of free electrons moving at the point.

As the electric field increases, this ionization process of corona current also increases to arc discharge, and a spark reaches out to meet the downcoming leader, forming a path for enormous currents to flow.

Radioactive lightning rods contain a certain amount of radioactivity at the tip of the rod, supposedly to enhance the ionization and, hence, attract the lightning leader over greater distances. These claims have been examined experimentally and theoretically by many scientists, with negative results. In effect, the analysis shows the corona current from the radioactive rod is slightly higher than that from the conventional rod, as the manufacturer claims, only when electric fields are low—such as under a fair sky. When a thunderhead approaches and the electric field builds up, however, the radioactive rod gives off less corona than the conventional rod and is, therefore, less likely to be struck. This can be explained by the fact that the ionization cloud produced around the rod by the radioactive source provides an ion shield at the tip, reducing its effectiveness in sending up the necessary leader spark.

The corona discharge from a conventional rod was found to exceed that from a radioactive rod by an order of magnitude under lightning-like electric fields, indicating that radioactive rods are much less capable of influencing the path of a lightning discharge than a conventional rod.

Comparative Test Results

The bases for bans on radioactive rods in several foreign countries were comparative tests, both atmospheric and laboratory based. Following are excerpts from published results of such experiments.

In a document entitled "Tests on Ionizing Rods," professors C. Bouquegneau, C. Gregoire, and J. Trecat, of the Faculte Polytechnique De Mons in Belgium, reported the results of their tests on radioactive and electrically excited rods.

On May 2, 1984, 80 negative strikes and 80 positive strikes were generated in the professors' laboratory. An alpha-radioactive rod fitted with the isotope Americium-241 was struck 40 times, and the identical rod without radioactive material was also struck 40 times.

On June 12, a beta-radioactive rod was struck by 41 of 80 positive strikes, and an identical rod which was not radioactive received 39 strikes. The same day, 40 negative strikes went to the beta-radioactive rod, and the competing rod, which was identical but not radioactive, was struck 42 times.

"Electrically excited" rods were tested against identical rods that were not electrically driven July 3. Approximately half of the 200 strikes generated went to the electrically excited rod, and the remainder to the passive rod.

The Belgian professors commented,

> Our tests show that ionizing rods, (radioactive alpha or beta, or corona) are completely inefficient; there is no influence on the probability of strikes at metric distances, then, a fortiori, at decametric and hectometric distances, when adding radioactive stuff to the lightning rods or when exciting them with a high positive potential as high as +25 kV (kilovolts).

In the limits of the allowed radioactivities on our roofs, a classical Franklin rod experiences exactly the same effect as a more sophisticated structure exciting some ionization completely inefficient to catch the stepped leader of the lightning discharge.

The Swedish Tests

One of the first reports on the inability of radioactive material to enhance the ability of air terminals to capture lightning flashes was presented by Harald Norinder, a professor at Uppsala University in Sweden. Professor Norinder conducted tests outdoors with radioactive and nonradioactive air terminals of identical configurations and in exactly similar positions. He very carefully measured their responses instrumentally under thunderstorm conditions and reported that they were identical.

Radioactive Rods in America

Radioactive lightning rods were sold and installed in the United States during the 1950s and 1960s, but with very limited success. Manufacturers and installers of code-adherent lightning protection systems were able to cite Professor Norinder's findings as well as comments by other atmospheric scientists. However, a few companies continue to market these devices. This effort has been somewhat successful due to a combination of aggressive marketing and a lack of data from tests conducted in the United States. The latter issue was recently addressed.

The American Tests

As the result of a history of periodic contact by phone and letter with Professor Emeritus Charles Moore, a highly respected atmospheric scientist at the New Mexico Institute of Mining and Technology, the author was aware of previous atmospheric testing of lightning protection air terminals in the lightning-prone mountains in the Socorro area. Professor Moore, now semi-retired, recommended that tests of radioactive lightning rods be conducted by a fellow atmospheric scientist, Dr. William Rison.

Dr. Rison received two such devices, known as "lightning preventers," from a U.S.-based vendor, along with special masts and installation hardware, in July of 1990. The equipment was installed in accordance with accompanying instructions during the period of August 9 through August 16. Dr. Rison wrote a lengthy report describing the experimental setup, the positions of the two devices (one contained radioactive material, while the other did not), the atmospheric conditions in the area, the effects of wind, and other factors.

Dr. Rison reported that "Robert Hignight, a Langmuir Lab employee, reported lightning within 100 feet of the preventers. That distance is much smaller than the 320-foot protective range claimed for the radioactive preventers." (See Fig. 17.1.) Had the vendor's claims of performance been fulfilled, the lightning flash reported by Hignight would have struck the lightning preventer that contained radioactive material.

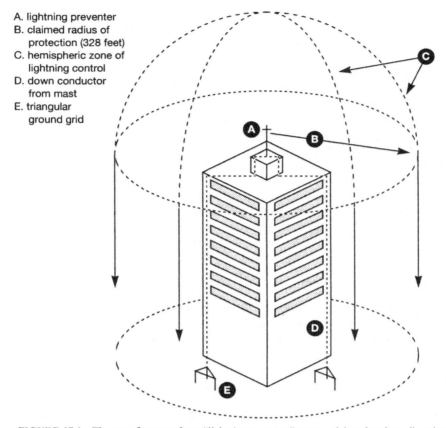

A. lightning preventer
B. claimed radius of
   protection (328 feet)
C. hemispheric zone of
   lightning control
D. down conductor
   from mast
E. triangular
   ground grid

FIGURE 17.1 The manufacturer of one "lightning preventer" system claims that the radioactive rod's protection umbrella extends over a radius of 328 feet. The true radius of protection is the same as that described in the code *NFPA 78* for a single, conventional rod of the same height.

In his report, Dr. Rison explained how and why radioactivity does nothing to improve the performance of an air terminal. Another, longer-duration test was conducted during the peak lightning storm period of 1991. Dr. Rison's report on that test described the test procedure, which was similar to the prior test's procedure and setup. He again reported no difference in the attractive qualities of the preventers, one of which contained radioactive isotopes and one of which did not.

Lightning Prevention Attempts

Nearly two and a half centuries ago, Benjamin Franklin speculated that sharply pointed lightning rods would "draw off the electric fire" from storm clouds and prevent lightning from striking to earth. It was not long, however, before the self-made atmospheric scientist modified his opinion, suggesting that if lightning rods failed to forestall a lightning strike, they would at least take the electric fire unto themselves and thus save a targeted building. While history is vague about Franklin's later conclusions, it seems likely that he eventually discarded the lightning prevention theory altogether in favor of the lightning capture and grounding functions that he had presumed and then observed.

However, "dissipation arrays" have recently been proposed as being effective in preventing lightning from striking a structure equipped with a multiplicity of sharp points. These arrays were once the prime subject of a discussion meeting of atmospheric scientists and other lightning and lightning protection experts, where the concept was roundly denounced.

The promoter of such arrays has steadfastly clung to his faith in them, and he argued forcefully for such arrays as alternatives to conventional systems during a meeting of the NFPA Lightning Protection Code Committee. The committee, however, voted unanimously to reject the system for inclusion in the code.

"Pulsing" Lightning Rods

In recent years, a French company introduced in the United States a system that applies a pulsed high voltage to the tip of a lightning rod of special design. The operational theory is basically identical to that of radioactive rods, and the claim is that artificial methods can be used successfully to boost ground-based electric charges upward more forcefully than can nature, thus greatly widening the protective area of a lightning protection air terminal.

There are two systems based on that premise. One employs solar panels and a battery to produce the oscillating voltage. A more recent innovation, a pointed lightning rod, is claimed to somehow draw energy from the surrounding electric field to propel the ground-based positive streamer upward with great force.

Artificially pulsed air terminals have not fared as well in the United States as have radioactive rods. The most likely reason for that is a difference in salesmanship and promotional ability.

Glossary

aerial energy source   generally, overhead electrical wires

ampere (A)   the basic unit of electrical current, equal to the motion of one coulomb of charge (or 6.24 x 10 electrons) past any conductor cross section in one second

atmosphere   a term designating an air pressure level that equals the force of that existing in a normal environment (approximately 14.7 pounds per square inch at sea level)

breakdown voltage   the voltage at which certain surge protection devices are engaged

capacitance   the property of a circuit element that permits it to store an electrical charge

cardiac arrest   stoppage of heart functions

cardiopulmonary standstill   stoppage of both heartbeat and breathing

clamping voltage   the voltage level at which a voltage-limiting device begins to provide protection

coalesce   to come together; unite

cold bolt   a brief duration lightning flash

comatose   unconscious

condenser   a device capable of holding an electric charge (e.g., a battery or capacitor)

convective storm   a storm where atmospheric conditions (e.g., heat) are transferred by massive motion

corona   a faint glow enveloping an electrode in a high-intensity field

corpusante   the glow atop a ship's mast during an electrical storm

coulomb   an electrical charge quantity transferred in one second by a steady current of one ampere

counterpoise   in this context, a grounding element that is able to distribute the electric charge evenly along its length; a buried surrounding cable

crimp-type   refers to connections formed by pressing or pinching connectors into folds or ridges

discrete   unconnected and separate

Dutch metal   a soft metal

electric field induction   a cause of relatively small current flows

electrode system   a grounding system of driven rods, buried cables, or other media

electrolytic interaction   a process producing chemical changes in metals, usually causing a degradation of conduction

electromagnetic field   a field of force consisting of magnetic energy in motion

electromotive force   the energy per unit charge that is converted into electrical energy from another energy source; e.g., chemical or mechanical

electron fluid   electrical flow

electrostatic potential   the strength of an electric field

electrostatic shielding   protecting conductors from invasion by overage current consequences

equipotential bonding   connecting two conductors so that they remain at the same electrical potential

exponential waveform   a waveform expressed in terms of a designated power

fibrillation   uncoordinated twitching of the ventricular heart muscle

firing voltage   the voltage at which certain surge protection devices are engaged

flash   the technical term for a lightning "bolt"

gently sloping roof   a roof having a span of 40 feet or less and a pitch of less than 1 foot
of rise to 8 feet of span, or with a span of more than 40 feet and a pitch of less than 1:4

gilt   a plating

harmonic vibration   sound waves caused by the outward thrust of air around a thunder-
bolt

hemispherical   consisting of half a sphere

hertz (Hz)   a unit of frequency equal to one cycle per second

high capacitance   the ability of a circuit to store a relatively large electrical charge

hot bolt   a long-duration lightning flash that applies heat long enough to ignite wood or
other materials

IC   as used in this book, an induction coil

induced voltage   voltage caused to flow along a conductor by a nearby flow

inductive consequences   the consequences of induced voltages

inductive heating   heating equipment using resistance of coils or wires to produce heat

inductive voltage   a voltage reduction caused by a reentrant loop

inductive   of or arising from induction

ionized   having been converted, totally or partially, into ions

ion   an atom or group of atoms or molecule that has acquired a net electrical charge by
gaining or losing electrons from a neutral charge

isokeraunic (or isoceraunic) level   the number of days annually that thunder is heard at
a weather station

kilohertz (kHz)   1,000 Hz

leader stroke   the initial flow of current in a lightning flash

magnetic field induction   a cause of relatively small current flows

magnetic field induction   the ability of a magnetic field to influence electrical potential
according to that field's magnitude and adjacency

magnetic field   a condition established in a certain space by a magnet or electric current
where a magnetic force is present

magnetic flux   the total number of magnetic lines of force passing through a bounded area
in a magnetic field

metal lattice   the arrangement of ions or molecules in metal

microsecond (μs)   one millionth of a second

middling wire   in this sense, a connecting wire

nanosecond   one billionth of a second

nitrate   a fertilizer; potassium nitrate or sodium nitrate

plasma   an electrically neutral, highly ionized gas compound of ions, electrons, and neutral particles

primary object   a metal object on or above roof level, outside a zone of protection, that can be struck directly (*see* secondary object)

Prometheus   in myth, the Titan who stole fire from Olympus

psychomotor functions   functions involving both muscular and mental activity

pulses   leader/return strokes in lightning flashes

reentry loop   a loop forming a part of a circle greater than 180 degrees

retrofit installations   installations made on existing structures

secondary object   a metal object that is inside a zone of protection but subject to a sideflash of current, depending on flash magnitude and adjacency

shock wave   the violent thrust of air outward from the flow channel of lightning current

shroud   a rope or wire from a ship's mast; a rigging line

solenoid   an electromechanical assembly often used as a switch, consisting of a coil and a metal core that is free to slide along the coil as influenced by a magnetic field

stroboscopic   the appearance of moving or vibrating objects remaining momentarily stationary

surge suppression   the act or technology of preventing damage, usually to electronic equipment, from harmful electric surge currents

tangent   making contact along a line, as in a horizontal grade line

tangential   moving along in the direction of a tangent

tort law   the law concerned with loss or casualty due to negligence

transfer resistance   the resistance to electrical charge conduction offered by a ground site

transient voltage   a temporary rise in current magnitude due to an external event

trigger   the process by which a tall structure carries ground currents upward to a point where those currents and oppositely charged atmospheric currents reach a point where they cause a breakdown of intervening air

ventricle standstill   stoppage of blood flow from the heart

vermiculite   a conductive compound related to chlorites, also helpful to plant growth

vortex   an aerial whirlpool

waveform   a mathematical representation of a wave of current

Index

CPSIA information can be obtained
at www.ICGtesting.com
Printed in the USA
LVOW04s1828090116
469918LV00005B/124/P

9 781468 465501